Praise for *Words*
in Water

"Sandy Rosenthal is a courageous and indefatigable warrior for justice."

—**Dave Eggers**, author of *A Heartbreaking Work of Staggering Genius*

"For years, Sandy Rosenthal has been the go-to source for any reporter seeking to understand the truth about flood protection in New Orleans: what really happened to the levees following Hurricane Katrina and the steps taken (and those that still need to happen) to protect the area and its residents. Read the inside story for yourself in this readable, engaging tale of how Rosenthal began investigating the US Army Corps of Engineers shortly after Katrina and her efforts to hold accountable the Corps and other public officials."

—**Gary Rivlin**, author of *Katrina: After the Flood*

"Putting herself in the path of the US Army Corps of Engineers (USACE), the local media, academia, and entrenched political interests in order to get to the truth about the New Orleans levee system took guts, as well as masterful community organizing. Anyone who is interested in Hurricane Katrina, and in America's failing infrastructure, will want to read this book told in a fast-paced narrative."

—**Scott G. Knowles**, head of the Department of History at Drexel University and author of *The Disaster Experts: Mastering Risk in Modern America*

"Sandy Rosenthal's account of the founding of levees.org after the flood of New Orleans is an invaluable memoir of the making of an activist. In a world crying out for citizen action around increasingly desperate climate issues, her story is timely and instructive—and even hopeful."

—**Michael Tisserand,** author of the award-winning *The Kingdom of Zydeco*

"Sandy is a New Orleans hero. Her advocacy on behalf of flood protection has changed this city for the better. My own book drew upon some of her research. Sandy can rally people to action like I've never seen and is relentless in her efforts to hold government and political leaders accountable."

—**James Cobb,** author *Flood of Lies: The St. Rita's Tragedy*

"For someone like me, who moved to New Orleans after Katrina, one of the biggest challenges of understanding New Orleans has been to understand what happened in that flood, and why, and what it means. I don't know of any individual who has done more to elucidate these matters than Sandy Rosenthal. Her scrupulous and ferocious focus on the facts has been a necessary tonic for the city and a help for its citizens. Sunshine is the best disinfectant and Sandy, in those terms, has been a huge source of light."

—**Thomas Beller,** associate professor of English and director of Creative Writing at Tulane, author, and contributor to the *New Yorker* and the *New York Times*

"An inspiring memoir and gripping detective story, *Words Whispered in Water* investigates the cause of the 2005 New Orleans flood in all of its muddied complexity. In this era of climate breakdown and failing infrastructure, Rosenthal's book is more than a history lesson. It's a master class in citizen advocacy and a rousing call to action."

—**Robert Verchick**, former EPA official in the Obama administration and author of *Facing Catastrophe*

"As I traveled through New Orleans in early January of 2006, there were no birds, no dogs, no children, no streetlights, no mail…but there were, I discovered as I drove around the haunted cityscape, yard signs in neighborhoods—seemingly everywhere—that said 'Hold the Corps Accountable.' They were, I soon learned, the work of Sandy Rosenthal, whose newly constructed Levees.org suspected, before two university investigations made it uncomfortably clear, that 'Katrina' was no natural disaster. Sandy is bright, dedicated, and fearless—a bad combination if you're an Army Corps of Engineers PR person.

"As those university investigations' results became known, at least to locals, Sandy turned her efforts to doing what would seem normal in other disasters but was ridiculously challenging locally—erecting monuments at key points of system failure to commemorate those who lost their homes, their livelihoods, or their lives through the mis- and mal-feasance of this mysterious federal super-agency. She has fought for her city, and her community, harder than most soldiers fight in war. And, with a new Corps-built system starting to reveal its own problems, she's not through fighting yet."

—**Harry Shearer**, actor, producer, and voice of *The Simpsons*

"For all the myths and misinformation that circulated after Hurricane Katrina, there are still people across the country who don't know that it was the collapse of the levees, not the hurricane itself, that drowned New Orleans. Sandy Rosenthal overcame hurricane hardships to establish, with a computer whiz of a son, the Levees.org website to get the story out and arouse the public. The lengths the Army Corps went through to silence her or deflect the truth and the stumbling blocks she experienced along the way are all here in a quick-moving tale. It makes for a gripping read."

—**Roberta Brandes Gratz**, award-winning journalist and author of *We're Still Here Ya Bastards*

"There are only a few civilians that fight like real warriors. Sandy Rosenthal is one of them. I've personally watched Rosenthal stand up for justice against big money companies with deep pockets. Even when the odds were stacked overwhelmingly against her, she came out the victor. Sandy's determined quest to put the truth about the massive flooding of New Orleans in 2005 in front of the American people is a story that needs to be told…and one that includes important lessons for holding powerful institutions accountable."

—**Russel L. Honoré**, Lieutenant General, United States Army (Ret.)

"Everyone alive at the time remembers the live-on-TV horror of Hurricane Katrina and the deadly drowning of a great city, New Orleans. The 'who, what, and why' of the catastrophe was thereafter carefully PR managed to protect the interests of the powerful, especially the notorious Army Corps of Engineers, shifting the blame to Nature and the city itself. It took a tireless, driven citizen movement to set the record right, confronting the failures of scientists and journalists as well as the devious smears, attacks, and propaganda of the powerful. Sandy Rosenthal led their fight and tells the story here. The truth of Hurricane Katrina has been terribly incomplete but is now unspun and revealed in this heroic book, *Words Whispered in Water.*"

—**John Stauber**, author of *Toxic Sludge is Good for You!*

Words *Whispered* in Water

Words *Whispered* in Water

Why the Levees Broke in Hurricane Katrina

By Sandy Rosenthal

Coral Gables

Published by Mango Publishing Group,
a division of Mango Media Inc.

Cover Design: M Styborski

Interior Design: Jermaine Lau

For permission requests, please contact the publisher at:
Mango Publishing Group
2850 S Douglas Road, 2nd Floor
Coral Gables, FL 33134 USA
info@mango.bz

For special orders, quantity sales, course adoptions and corporate sales, please email the publisher at sales@mango.bz. For trade and wholesale sales, please contact Ingram Publisher Services at customer.service@ingramcontent.com or +1.800.509.4887.

Words Whispered in Water: Why the Levees Broke in Hurricane Katrina

Library of Congress Cataloging-in-Publication number:

ISBNs: (p) 978-1-64250-327-2 (e) 978-1-64250-328-9

BISAC: SOC040000, SOCIAL SCIENCE / Disasters & Disaster Relief

Printed in the United States of America

For Steve, Aliisa, Mark, and Stanford

because our house was the levee channel,
twenty-four hours a day, seven days a week.

Table of Contents

Major Levee Studies

Study Name	Nickname	Date
Independent Levee Investigation Team (Raymond Seed, Robert Bea, J. David Rogers, et al.)	Berkeley team	June 2006
Team Louisiana (Ivor Van Heerden)	Team Louisiana	March 2007
"The New Orleans Hurricane Protection System: What Went Wrong and Why: A Report by the American Society of Civil Engineers External Review Panel" (C. F. Andersen, J. A. Battjes, D. E. Daniel, et al.)	ERP report	June 2007
"Decision-Making Chronology for the Lake Pontchartrain and Vicinity Hurricane Protection Project" (Douglas Woolley and Leonard Shabman)	Woolley Shabman paper	March 2008
Interagency Performance Evaluation Taskforce (Lewis E. Link, John Jaeger, Jeremy Stevenson, et al.)	IPET	June 2009
"Interaction between the US Army Corps of Engineers and the Orleans Levee Board preceding the drainage canal wall failures and catastrophic flooding of New Orleans in 2005" (J. David Rogers, Paul Kemp, H. J. Bosworth Jr., and Raymond Seed)	*Water Policy* paper	March 2015

Prologue

Just before the eye passed east of New Orleans, a hurricane surge entered the 17th Street Canal, the largest drainage canal in the city. Floodwalls groaned against the surge's weight despite supporting steel pilings anchored into thick, earthen levees. The mighty 17th Street Canal could move nearly 10,000 cubic feet of water per second, enough to drain an Olympic-size swimming pool every nine seconds.

But, on this particular Monday morning (August 29, 2005), something was wrong. A section of the floodwall atop the levee had begun to tilt. The steel pilings were too short, and water was flowing into the open gap. The entire section of the floodwall and the levee slid sideways, unleashing a furious blast of briny water into the nearby (mainly white) neighborhood of homeowners.

Eighteen years earlier, the Army Corps of Engineers (Army Corps) had decided that driving steel pilings deeper than sixteen feet was a waste of money.[1] Originally, the design for the canal's proposed new floodwalls had called for expensive steel pilings driven forty-six feet into the ground. But the agency was behind schedule, and costs were rising. In response, the Army Corps conducted a large-scale test study to find ways to save money on steel.

Tragically, they missed a warning sign.

During the test study, when the steel pilings were subjected to a test water surge, they had tilted. No one noticed the menacing tilt because the pilings were underneath a tarp. As a result, the engineers determined that they needed to drive down the steel pilings only sixteen feet instead of forty-six. The Army Corps used this alternate engineering rule for new floodwalls on the 17th Street Canal and several other canals across the city. The new rule saved the Army Corps a total of $100 million.[2]

In 2000, new floodwalls were installed, but they were destined to fail.

When they collapsed five years later—at a fraction of the water pressure they were designed to contain—hundreds died instantly and thousands more died within months. New Orleans was devastated to the tune of well over twenty-seven billion dollars because floodwalls were not correctly designed and built by the Army Corps.[3]

At 7:08 p.m., EST the day after the floodwalls broke (August 30, 2005), the Army Corps went into full-time, damage-control mode. Its spokespersons told big media outlets that the hurricane storm surge was just too great. Water had flowed over the 17th Street Canal's floodwall and caused it to collapse.[4] Their story over the course of the next two years would be that nature caused the destruction of New Orleans and New Orleanians themselves were responsible for their loss and suffering due to their own stupidity (they live below sea level) and sloth (the local levee officials were lazy). The Army Corps' primary mission had become rewriting history and duping the American people.

They almost got away with it.

Chapter

1

Goodbye, New Orleans

"Pack like you're not coming back home," barked my husband Steve to our college-aged son.

"Dad, that's a terrible way to wake me up!" grumbled Mark, startled by this grim command.

It was 6:00 a.m. on August 28, 2005—the day before a major hurricane was forecast to strike the southeast coast of Louisiana. Two hours earlier, the National Weather Service (NWS) had upgraded the storm to Category 5, the highest measure on the Saffir–Simpson scale. After rousting the rest of us none too gently, Steve explained that this was a bad one, and he did not think we could get back to our house sooner than three weeks.

"Mark, pack everything you need right now to go straight back to college." Then, to all of us, he said, "We're out the door in an hour."

Steve, Mark, and our fifteen-year-old son Stanford had spent the prior day boarding up our two-story house under a blazing sun. In the eighty-eight-degree heat, they put plywood on every window of our uptown New Orleans home. Category 5 winds are over 154 miles per hour, and we prepared for them.

Born and raised in this historic city, my husband knew well the damage the storm's winds could inflict. Forty years earlier, Hurricane Betsy had ploughed through the city with winds up to 110 miles per hour. In 1965, the technology to provide early warnings had not yet been invented. By the time the city's residents knew that a hurricane was churning toward them, it was too late to evacuate. Steve's family did what everyone else did: they hunkered down and weathered the storm. All night long, the winds battered the house with a deafening noise. At daybreak the next morning, Steve recalled, he pried open the front door and saw that all the leaves had been stripped from the trees and had plastered every house, building, and structure with a new coat of bright green. It was a breathtaking memory. But he also remembered having no electricity, no air conditioning, and no ice for six weeks.

Born and raised in Massachusetts, I had never experienced the childhood excitement of a hurricane passing over my house, but I had heard all about it in my twenty-seven years of marriage. I understood that if the oncoming hurricane was slow-moving and dropped a lot of rain, we could have some water in the house. Natural river levees to the south and natural lake levees to the north team up with man-made canal levees to the east and west. If the rain fell too fast, the pump stations could not keep up. So while the menfolk were boarding up the house, I was inside doing the lighter work of moving furniture and curtains out of harm's way. We were fortunate to have two stories.

* * *

However, the Millers were not as fortunate. Nine miles away in the Lake Vista neighborhood of New Orleans, Harvey and Renee Miller decided to ride out the storm in their home. After all, they had ridden out at least five or six storms. And besides, they had a safe harbor if needed: keys to an empty two-story house only two doors away. Their own home, and every other house in the neighborhood, was single story.

* * *

After boarding up the exterior, Steve went to the front yard and plucked ten almost-ripe papayas from our trees. They would be good eating in the days ahead when we would be living mainly on unhealthy fast food. Though Steve was convinced that we would not be able to come home for three weeks, there was a chance that he may be wrong. So, in the kitchen, we "bought time" by placing four large pots of water in the freezer. Just before departing for our evacuation destination, we would divide the pots between the freezer and refrigerator to create vestiges of old-fashioned ice boxes. The slowly melting chunks of ice would help keep our food from perishing for up to three days without electricity.

A habitual list maker, I adhered to the same hurricane evacuation list I had used just one year earlier for Hurricane Ivan. In advance of that 2004 storm, Governor Kathleen Blanco had ordered a mandatory evacuation for all the coastal parishes of southeast Louisiana and a voluntary evacuation for New Orleans. With the discomfort of remaining in New Orleans for Hurricane Georges in 1998 still fresh in our minds, we had opted for the luxury of electricity by evacuating to a motel in Jackson, Mississippi.

Hurricane Ivan wasn't remembered for its ferocity; rather, the name became synonymous with state government incompetence. The year 2004 was the first time that Louisiana had employed "contraflow," turning all lanes of the interstate into one-way roads outbound from the city. The concept is simple, but the devil is in the details. The system requires perfect coordination between local parish governments and the state police. That very first use of contraflow to evacuate residents for Ivan was fraught with problems, mainly due to the unexpectedly large amount of staff required to close highway exits. Traffic delays were so severe that many turned around and went back home in time to watch Ivan sputter and weaken to a tropical depression.

* * *

Harvey Miller filled his car with gas. Then he and his wife Renee brought food and a couple of gallons of water to their neighbor's two-story house. They also brought a cot, folding chairs, and a portable television, most of which they left on the first floor. The house was raised, meaning that it was ten full steps to reach the first floor. The water might rise to the front door, they thought, but no higher. With everything ready and in place, they settled down for the evening in their own home, believing that their "safe harbor" would make everything alright if needed.

* * *

By the prior Wednesday (August 24, 2005), my husband had already reserved twenty motel rooms in Jackson, Mississippi, three hours due north. Two rooms were for our family; the others were for the families of critical employees with Strategic Comp, my husband's workers' compensation insurance company. The Days Inn, which had been previously scoped out and selected, had reliable internet connections and was pet friendly. Reserving rooms five days in advance for a possible hurricane might appear overly cautious to most, but the arrangements can be made in five minutes and canceled in one.

During the day on Friday, August 26, some of the computer models shifted the track of the storm, now a hurricane, west. Then, late in the afternoon, the models shifted in unison, and New Orleans was moved to the center of the cone of certainty.[5] Governor Blanco declared a state of emergency at four o'clock. Mayor Ray Nagin followed suit but stopped short of calling for a mandatory evacuation. It was now certain that we would need those motel rooms.

By 7:00 a.m. on Saturday (August 27), the hurricane was over the center of the Gulf of Mexico. At first, the eye started to disintegrate, normally a sign of weakening, but in this case it was redistributing. Wind speed picked up around the central vortex, and pressure fell again. Later, the eye contracted, and masses of thunderstorms sprang to life. Within a few hours, the storm doubled in size, eclipsing most of the gulf.

Throughout the day on Saturday, radio and television reports urged residents to evacuate. Officials for Plaquemines and St. Charles Parishes (low-lying coastal areas south of New Orleans) ordered mandatory evacuations. The governor ordered contraflow to be put in effect, and by four o'clock that afternoon the state police had reversed all inbound lanes. By this time, the hurricane watch had been widened to include everything from western Louisiana to the Alabama-Florida border.

The National Hurricane Center (NHC) issued a bulletin that warned of a powerful hurricane with unprecedented strength: "Most of the area will be uninhabitable for weeks. Perhaps longer. At least one-half of well-constructed homes will have roof and wall failure. All gabled roofs will fail… Water shortages will make human suffering incredible by modern standards."[6]

Then, on Saturday night, Max Mayfield, then director of the NHC, did something he rarely did. He called all the governors in the cone of certainty to warn them. Upon urging from Governor Blanco, he also called Mayor Nagin, telling him that some levees in the Greater New Orleans area could be overtopped."[7]

* * *

At nine o'clock on Sunday morning (August 28)—the day before the levees broke—we parked my three-year-old sedan on our elevated driveway and climbed into our packed Ford Expedition to depart for Jackson. Just before leaving, we checked on our elderly neighbors because we were worried about them. Steve had spoken to Charles Prince the day before and, at that time, he and his wife Zelda planned to shelter in place. Now that the storm had swelled to a Category 5, staying was not an option for our neighbors.

"Charles, I've been through a hurricane like this," my husband told him. "You won't have electricity for a month. And that's not all. All these big trees will come down, and you won't be able to drive your car."

"I'm a World War II veteran, and I've been through hurricanes," Charles answered defiantly.

Charles and Zelda had moved to New Orleans from Long Island, New York, fifteen years earlier and had little comprehension of the danger they faced. Steve realized that he had no choice but to frighten Charles into action.

"What if something happens to Zelda? If there's an emergency, you won't be able to get her medical treatment."

That convinced Charles, but he claimed that he had no family within driving distance, and it was too late to find a motel room. We offered to give them the motel room that we had reserved for our two sons, and they would sleep on the floor in our room until we got things figured out. I would find out later that hundreds of elderly residents refused to leave their homes. Many stayed to care for their pets, but most were just too stubborn.

* * *

It was time to go. We backed out of our driveway after Charles and Zelda got into their car to follow us to Jackson. Mark and

Stanford were jammed in the back seat between suitcases and bags. Our tiny dachshund Chester jumped back and forth from the back seat to the front in his usual overexcited state whenever we took him for a car ride. (My twenty-two-year-old daughter Aliisa was living and working in New York City, having just graduated from Brown University.)

We drove down elegant and picturesque St. Charles Avenue toward the contraflow evacuation route. The stately Southern homes were eerily quiet as most people had departed the previous day or earlier that morning. House after magnificent house was boarded up. The proverbial hatches were battened down.

At nine thirty, we turned on the radio and heard Ray Nagin, our soon-to-be-infamous mayor, describe the first ever mandatory evacuation[8] of New Orleans, which state police had called at 8:17 a.m. Though his voice was calm, he implored the city's residents to leave. Max Mayfield's warning had apparently frightened the mayor. "You need to be scared. You need to be concerned. And you need to get your butts moving out of New Orleans right now."

I listened raptly to the radio. Despite the mayor's tranquil tone, his message was blood-chilling.

"When the floodwaters rise to your second floor, you will need an ax to chop a hole in the roof," he admonished.

"Goodbye, New Orleans," I murmured to the passing houses.

I leaned down to rearrange my tennis gear on the floor of the cramped car to give my feet more room. Just before leaving the house, I had decided to bring my tennis gear. Oddly, because of that decision, I would soon challenge one of the most powerful bureaucracies in the world.

Cars streamed out of the city all day. Under the contraflow plan, on-ramps worked, but exit ramps did not. There was nowhere to go but out. The key to contraflow is the phased evacuations, fifty hours in advance in Louisiana's southern-most parishes, and thirty hours in advance for New Orleans. With the improved contraflow system, our journey to Jackson was relatively painless.

Years later, it was agreed upon that Hurricane Ivan saved a lot of lives on August 29, 2005, because that earlier storm exposed a contraflow plan in need of revision. If not for the dry run of Hurricane Ivan in 2004, thousands might have abandoned their attempt to evacuate for the 2005 hurricane. Governor Blanco's evacuation of 90 percent of the New Orleans region in 2005 using contraflow would be cited as the most successful rapid evacuation of a major city in American history.[9] Nonetheless, the trip took over twice the time it would have taken in sunnier weather.

In a noon teleconference on Sunday (August 28), Mayfield said, "On the forecast track, if it maintains intensity, about twelve feet of storm surge in the lake, the big question is will that top some of the levees? … I don't think any model can tell you with any confidence right now whether the levees will be topped or not, but that's obviously a very, very grave concern."[10] While talk of possible overtopping was discussed right up to the first storm surge, there was no warning that the levees could break.

* * *

We pulled up at about four o'clock to the small, two-story motel about a mile from Jackson's city center. The motel was full to bursting with New Orleanians and their pets, but the mood was upbeat. Smiles were abundant, and the important work of caring for the needs of pets was a good distraction. And besides, most thought they would be going home in two days: everyone except the Rosenthals. We planned to be in Jackson for at least three weeks. I unpacked our suitcases of clothes, supplies, and equipment while Steve checked on his employees to make certain that everyone had their laptops for work and that the internet was accessible and functioning.

I touched base with the Jacobs family. Steve's sister Leslie, her husband Scott, their daughter Michelle, and their dog Cayenne had two rooms on the second floor alongside ninety-three-year-old Rose Brener (Grandma Rose), my husband's maternal grandmother, who needed special care. Grandma Rose was an amazing lady. The family called her "Grand Central" because she remembered every detail of everyone's life. Grandma Rose was not exactly happy to be living in a motel room evacuated for a hurricane; however, she was a model of strength, having survived many catastrophes including the Great Flu Epidemic of 1918–1919 in New Orleans and, of course, the Great Depression from 1929 to 1939. Later that day, Leslie and Michelle departed for Philadelphia on the last flight out of Jackson. Michelle would be starting her first year at the Wharton School of the University of Pennsylvania. Scott stayed behind to take care of Grandma Rose and Cayenne.

Around seven thirty, we hopped into our car and drove to our favorite Jackson eatery—a family-owned Thai restaurant that we had visited the previous year during our evacuation for Hurricane Ivan. This time, the owner welcomed us with special delight because he remembered our gift to him the year before: one of our papayas. Later, back in our room with tummies full of chilled, spicy green papaya salad and warm duck curry, the whole family watched television for developing details of the storm. Its strongest winds were blowing at about 175 miles an hour, and the center was 200 miles from the mouth of the Mississippi River.[11]

By early evening, the state police suspended contraflow. Most of the metropolitan population of one million people had left. But about 100,000 people remained inside the city, most of them in their homes, and about 14,000 people had taken refuge in the Superdome.[12] It was time to hunker down and ride it out.

* * *

While my family and I slept, emails zinged back and forth between the NWS, the Federal Emergency Management Agency (FEMA), and other agencies. One email was focused on storm surge: "Any storm rated Category 4 on the Saffir–Simpson scale will likely lead to severe flooding and/or levee breaching. This could leave the New Orleans metro area submerged for weeks or months."[13] This email, sent long after the evacuation was completed, was the first to suggest the possibility of levee breaching. It went up the chain of command. At 1:47 a.m., the Homeland Security Operations Center watch officer emailed it to the White House Situation Room.[14]

The events of the next few hours are the reason that the world remembers this day: August 29, 2005.

At about five in the morning, a thirty-foot section of floodwall—called a "monolith"—on the east side of the gigantic Inner Harbor Navigation Canal (known locally as the Industrial Canal) breached and released stormwater into the adjacent Lower Ninth Ward, a dense neighborhood of primarily black homeowners. The breach spread to 250 feet wide,[15] next to the blue-painted Florida Avenue bridge in front of the New Orleans Sewerage & Water Board (S&WB) pump station 5. Operators throughout the metropolitan area listened to their brethren beg for help as the station flooded.

* * *

Elsewhere in the city, there were other signs that something sinister was unfolding. Harvey and Renee got up at seven o'clock to foot-deep water all around their house. Without hesitation, they rounded up their black Labrador, Monet, and the trio sloshed through the water to what they expected would be safety and relative comfort. But it turned out that both the water and the gas were turned off. At the very least, they were able to turn on the battery-operated television. On the news, they saw that the marina was burning.

* * *

At approximately 7:45 a.m., a much larger second hole opened up in the Industrial Canal just south of the initial breach. The horrific surge of water lifted up whole homes nearly intact, which all joined a ghastly parade in the flowing current on Tennessee Street before disintegrating. Floodwaters from the two breaches combined to submerge the city's entire historic Lower Ninth Ward neighborhood in over ten feet of water.[16]

Levee monoliths on the west side of the Industrial Canal also gave way and flooded the Upper Ninth Ward, Bywater, and Treme neighborhoods in water between four and six feet deep.[17] Storm surge overtopped and melted levees that should have protected the primarily residential northeastern portion of New Orleans (locally called New Orleans East) and put that entire area underwater.

Farther to the west, between six and seven in the morning, a monolith on the east side of the London Avenue Canal failed and allowed water over ten feet deep into the mostly black Fillmore Gardens neighborhood.[18] At about six thirty, on the western edge of the city, several monoliths failed on the 17th Street Canal.[19] A torrent of water blasted into the mainly white Lakeview neighborhood of homeowners. Several firefighters, who had sheltered in the upper floors of the Mariner's Cove condominium complex, watched in shock as water flowed through the breach to a depth of over ten feet.[20]

The final breach of the flooding catastrophe took place between seven and eight in the morning when the west side of the London Avenue Canal breached—in addition to the east side—and flooded the mixed-race neighborhood of Lake Vista.[21] This second breach is a startling testament to the walls' weakness.

* * *

Harvey and Renee Miller did not hear this final breach just two city blocks away from them, but Renee was the first to notice dark spots on the wall-to-wall carpeting. Water was coming up through the floorboards. They rushed to the window and watched in disbelief as water surged down the street like river rapids. They stood stunned and watched as—in a matter of seconds—a Cadillac was picked up, carried down the road, and deposited in a tree.

But there was no time to watch because cold water was already a foot deep in the house, and they had to get upstairs! Monet balked at first because she had never seen steps inside of a house, but they managed to coax her up. They brought everything they could up to the second floor—the food, fresh water, and the television set—as floodwater rose shockingly fast inside the house. Fifteen minutes later, on Harvey's last trip downstairs, he found that the water came up to his chest. At that moment, he understood that their own home must be underwater. Exactly one month earlier, Harvey and Renee had made the final payment on their home.

* * *

As we ate breakfast in blissful ignorance at the Days Inn, the City of New Orleans was going underwater. The catastrophic breaches of the Industrial Canal, the 17th Street Canal, and the London Avenue Canal immediately killed hundreds and destroyed homes, commercial buildings, and infrastructure.[22] Breaches of levees and floodwalls on two other navigation canals—the Mississippi River–Gulf Outlet (MR-GO) Canal and the Gulf Intracoastal Waterway (GIWW)—were also releasing floodwater that would inundate the rest of New Orleans and nearby St. Bernard Parish. Police stations began flooding as storm surge sliced up the communications webs on which emergency managers in Baton Rouge and Washington, DC, relied. Landlines and switching stations were being submerged, and the 911 system went down. One by one, police, city, state, and federal agencies were blinded.[23]

* * *

Still oblivious at our mid-state Mississippi motel, we passed some of the time with Grandma Rose doing a crossword puzzle. We wandered through the motel, making friends and tending to our pets. We checked on Charles and Zelda but found out that they had departed at sunrise. They left a note, explaining that they had decided to drive to Zelda's sister's home in Birmingham, Alabama.

Jackson was a reasonable choice as an evacuation destination. But, in the case of this particular hurricane, the motel was now partially in the storm's path. Around noon, the hurricane's outer bands had begun sweeping through Jackson, bringing winds over fifty miles per hour and accompanying rain. The storm veered eastward and eventually lost strength. But that evening, the winds knocked out our motel's power. Only the fire escapes were lit.

The managers of a nearby Baskin-Robbins restaurant brought all their ice cream to the motel to be enjoyed rather than watch it spoil. While we ate our melting ice cream, FEMA director Michael Brown painted a grim picture of a ruined city on CNN's *Larry King Live*. But he was coy about discussing levee breaches: "We have some, I'm not going to call them breaches, but we have some areas where the lake and the rivers are continuing to spill over."[24]

At the Homeland Security Operations Center, Matthew Broderick, the marine brigadier general in command, received reports that levees were breaching, but there was also a televised image of people in the French Quarter toasting their survival on high, dry land.

Broderick concluded that he needed to have more specific detailed information in hand before reporting to the president of the United States. And he went home to bed.[25]

* * *

With nothing to do but fret in the dark, we decided to go to bed too. Normally, a Mississippi night in late August with no electricity would be insufferable, but the wind and rain had cooled the outdoor temperature to a reasonable seventy-two degrees. We fell into an uneasy sleep along with hundreds of thousands of other evacuees.

The next morning, amid downed trees and power lines, it looked like electricity would not be restored for days. The whole point of evacuating to Jackson was so that my husband could continue operating his workers' compensation company. This meant that we, along with his other employees at the motel, needed to pack up and relocate. We decided on Lafayette, Louisiana, which had not lost power and had a large, dog-friendly hotel: the Drury Inn, just off Interstate 10 (I-10).

So, for the second time in two days, our extended family evacuated. For the entire ride, our ears were tuned to Garland Robinette on WWL (AM) radio, trying to learn what was happening in New Orleans. We knew that power was knocked out, but other strange reports were trickling in. We heard that thousands were still in the city at the Superdome and that no help was in sight.

* * *

At 7:08 p.m. EST, the Army Corps issued its first press release, stating that the agency believed the 17th Street Canal had overtopped, which caused its collapse—a claim immediately proven false by eyewitnesses and later by multiple levee investigation teams. The press release ended with this statement: "The New Orleans District's 350 miles of hurricane levee were built to withstand a fast-moving Category 3 storm. The fact that Katrina, a Category 4-plus hurricane, didn't cause more damage is a testament to the structural integrity of the hurricane levee protection system."[26]

It is possible that this initial information was off-base due to the environment of sheer chaos caused by land-based and cellular phones having become useless in Greater New Orleans proper.

* * *

Renee and Harvey Miller were initially panic stricken, but by Tuesday afternoon (August 30), they realized that they were safe where they were. They had food, water, and a couch. There was no water for the toilet, but they had a bucket and plenty of floodwater—all the comforts of home except for just one thing: the dog had to go out! For this challenge, they managed to get their very accommodating Monet to use the roof for her business. Off and on throughout the day, they heard helicopters flying overhead. Each time, they went to the windows and waved frantically, but no one paid attention.

* * *

Michael Chertoff, secretary of Homeland Security, saw no reason to interrupt his trip from Washington, DC, to Atlanta to discuss the threat of avian flu with the Secretary of Health and Human Services. But, at the meeting's conclusion at noon on Tuesday, August 30, he was dismayed to learn that things in New Orleans were spinning out of control. He decided that the situation needed his full attention and, upon returning to Washington, he drafted language that invoked the National Response Plan.[27]

Meanwhile, engineers from the Army Corps and staffers with the Orleans Levee District and state Department of Transportation and Development all met at the 17th Street Canal breach site. They realized that none of the agencies had an emergency contingency plan for sealing levee breaches.[28] Such an absence of planning for breaches spoke to the fact that, in 2005, the Army Corps was considered the gold standard in levee-building.

* * *

We arrived in Lafayette at about four o'clock on Tuesday afternoon (August 30), at the jam-packed, four-story hotel. The mood at the Drury Inn was far different from the Days Inn in Jackson just twenty-four hours earlier. A forced, three-day mini-vacation and a "let's make the best of this" attitude had turned into a bona fide evacuation with fear and anxiety. One small blessing in this hotel was that our two boys had a room of their own, and Steve and I were lucky enough to get one of the hotel's few rooms with a tiny kitchenette.

After waiting over an hour for the single, coin-operated washer and dryer, we drove to Shoney's for some dinner. We spoke little. Our eyes and everyone else's were fixated on the television high in a corner of the busy restaurant. There was nothing new, only the same reports over and over. We drove back to the Drury Inn and climbed the stairs to the fourth floor. Walking down the hall, we could hear the sharp bark of our very yappy wiener dog.

Steve and I settled down with Grandma Rose and Scott to watch television, hoping for something different from the hurricane-racked region. Then we got the most shocking news of our lives. The levees had broken, the "bowl" was filling, and the City of New Orleans was going underwater. And that wasn't all; stories told of chaos at the convention center, looters at One Canal Place shopping mall, and "marauding black youths" walking down St. Charles Avenue and breaking into each house, one by one.

We did not know it yet, but our house—six feet above sea level along the Mississippi River—was high and dry. We also did not know that the reports of looters were exaggerated. Nonetheless, the night of August 30 was tearful for everyone. We all thought of the city and those still in New Orleans. Would the city come back? How many had died?

Steve and I walked from Grandma Rose's hotel room on the third floor to our sons' shared room on the second floor to tell them this terrible news. But they already knew because they were watching the same news reports. Of course they were. For a while, we just sat together, saying nothing. After asking them if there was anything they needed, we said good night, and shuffled back to our room. Unlike when we arrived in Jackson, I was too numb to unpack, too numb to do anything except watch the news reports on CNN over and over before falling into the first of countless uneasy nights' sleep.

Chapter

2

The Flood

Early in the morning on August 31, 2005—two days after the floodwalls broke—Harvey Miller awoke to someone yelling in front of the house. It was a man in a tiny canoe, paddling with a kitchen broom. He told Harvey that two neighbors a few houses away had chopped their way out of their attic and that he had canoed them both to safety. The neighbors had made him promise to return and get the Millers. Harvey eyed the tiny canoe. Renee did not know how to swim, and this looked just too dangerous. He asked Canoe-man if he could send someone back with a larger boat. Reluctantly, Harvey watched him broom-paddle away.

* * *

On most mornings before the 2005 flood, I woke up slowly, cherishing that warm, comfortable place next to my husband. But on Wednesday (August 31), in a hard bed at the Drury Inn, I snapped to full-blown awake as though someone had thrown ice water on me. I would wake up like that every day in the coming months.

Few people can viscerally comprehend surviving a catastrophe that claims not just an entire neighborhood but hundreds of neighborhoods in the space of a few hours. The scope of the disaster was now becoming apparent as images of men, women, and children standing on rooftops flashed across the videosphere. These people had lost much more than their homes and all their contents. They also lost their places of worship, their favorite stores, their doctors' offices, and their hospital. Many lost a car. They might have lost the plants and foliage they loved to tend that were destroyed in the briny gulf water.

The 2005 flood also affected their parents, their sisters, and their brothers. Too many had lost loved ones. When the counting was over, the catastrophe had claimed the lives of 1,577 people in Louisiana, including 1,300 directly due to flooding or wind according to the National Oceanic and Atmospheric Administration (NOAA).[29] The number rises to 2,000 when you include the trauma of relocation, illness, and suicide.[30]

Everyone who got through the trauma managed it differently. Some worked nonstop; some cried nonstop. During these dark days, I leaped out of bed because of the long list of things that needed to be done. What about people who were depending on me? I was now living in a different city. Every waking moment, I focused on caring for the needs of my two sons and little dog which pulled my attention away from the radio's depressing (and often wrong) information slowing oozing out of the city. It would take months before anyone knew the full scope of the flood disaster, and it would be years before those responsible would be identified. But on Wednesday—two days after the floodwalls broke—we clung together, often silently soaking up every bit of news we could find. None of it was encouraging.

The storm had passed north of the city, and surge levels had dropped, but water continued to flow through breached levees fed by the swollen Lake Pontchartrain just to the north. Water also flowed into neighboring suburban Jefferson Parish to the east. Floodwaters equalized with the lake and reached their highest point around midday on Wednesday. The news reported that it would be months before New Orleans could be drained—another falsehood. Responsibility for the lack of reliable information falls squarely on FEMA, and it may have been FEMA's single most egregious crime. Here's why.

A FEMA team, led by Phil Parr, had helicoptered into the Superdome at noon on Tuesday (August 30). Parr had expected to work out of Red October—a high-tech, mobile communications center with thirty computer workstations on the back of a tractor trailer. Its satellite phones and internet capability could have provided a makeshift network and allowed first responders to communicate, something literally worth life and death.[31] But Red October was still a six-hour drive away in Shreveport because two people had ordered it—Phil Parr and Michael Brown—and the request was countermanded by FEMA headquarters.[32] Now it was too late to get it to the Superdome due to the water and debris, and Parr's job was irrelevant. Imagine how different it could have been with Red October for police and first-responder rescue teams! So many people could have been saved. And it would have damped many of the rampant, ridiculous rumors.

The summer of 2005 should be the last time in history that an entire metropolitan area is cut off from the rest of humanity.

* * *

In this unreal world, Renee and Harvey were still trapped in their two-story safe house. Several hours after Harvey had declined help from Canoe-man, Harvey became angry. For days now, he had been watching while his neighborhood disintegrated and their clean water supply ran low.

“This is ridiculous!” he said aloud. “Why is no one coming for us?”

So he stripped to his shorts, descended the stairs into the cold water, and swam out the front door. Large, speckled trout surrounded him and traveled alongside him as he breast-stroked to his house on Charlotte Drive. Harvey had to see for himself if everything in his home was gone.

Then he saw a helicopter and started to splash and yell, “HELP!”

The helicopter’s strong wake threatened to drown him, so he swam to an oak tree and wrapped his arms around a thick branch. The helicopter flew away.

Clinging to the tree, Harvey shook his fist and yelled, “Why is this happening to me?”

Months later, on a trip back to the city, Harvey would see that the oak branch, to which he had clung three days after the floodwall broke, was easily twelve feet above the ground.

* * *

As Wednesday morning became afternoon, the Rosenthal family continued to do what every evacuated family did: focus on the most basic things in life. Where will we work? Where will the children go to school? And, after those first two things were decided, where will we live?

Our first priority was finding office space for my husband’s company to operate, which also included me. I did the outreach marketing for the company with a specialty in copywriting. We still did not know the condition of the office on Edenborn Avenue in the New Orleans suburb of Metairie. Nor did we know which employees had flooded homes. But even if the Edenborn office was not flooded or wind-damaged, few people would be living in the Greater New Orleans region.

Steve dispatched his office manager to find space in Baton Rouge. Moving quickly was critical. Other businesses were doing the same thing by moving their operations temporarily to Baton Rouge, to Lafayette, and to the north shore of Lake Pontchartrain. Waiting even a day or two could mean finding nothing at all.

With quick wit and fast driving on I-10, Steve managed to find office space at 5700 Government Boulevard. The building was a bit run down, but we got an excellent price and the space fit our needs. We would move in twenty employees as soon as furniture, equipment, and supplies were delivered—compliments of the Great American Insurance Company, the joint-venture partner with my husband's company. They had come to the rescue by providing these things gratis. The plan was to open on Wednesday, September 13.

During the one-hour drive back to our hotel room in Lafayette, we tuned in yet again to Garland Robinette on WWL (AM) radio in New Orleans. "Garland," a well-known, trusted voice, was filling in for a sick friend when he went on the air on August 29, 2005. Later, due to the station's strong signal, Garland was dubbed "the voice of New Orleans" because his was practically the only one heard during that interminable week. On this day, we found out that the Coast Guard's search for people in immediate danger was still ongoing, and there was no contingency plan for people who were in relative safety but trapped in appalling conditions.[33]

We arrived back at the Drury Inn just as two busloads were pulling in from disaster zones. Most were elderly residents of south Louisiana, and each had a story more harrowing than the next. It seemed that there were two predominant reasons why aging residents had chosen against evacuation: they had never flooded before or they had to care for their pets—or both. The stories of howling dogs being peeled from their crying owners' arms by rescue workers haunts me to this day.

* * *

Harvey Miller clung to the oak branch and started to further despair as he saw the helicopter fly away. But then he saw that a Coast Guardsman had been let out of the helicopter onto a nearby roof.

He yelled to Harvey, "We can bring a boat for you and your wife!"

As Harvey breast-stroked back to the house, a forty-foot motorboat towing a pirogue (a long narrow canoe) pulled up. Two big, burly men helped Renee and Monet through the window and into the boat. Harvey was too big to fit through the window, so he went downstairs, and the boat pulled up to the front door.

In that incongruous moment, with the helicopter chugging overhead and a boat backing away from the house, Renee called out, "Did you lock the front door?"

Everyone in the boat laughed.

* * *

On Thursday morning (September 1), we woke up in our insular world. Our cell phones were charged but useless. Email communication was effective but only for those who used it. Therefore, I was still not able to contact my family in New England. I could only imagine what horrors they envisioned if they were watching television. I needed to tell my family that we were all fine and that we evacuated with clothes and supplies for several weeks.

On this morning—three days after the floodwalls broke—in a world unto ourselves, we at least knew where Steve and I were going to work. Now we needed to take care of the second order of business. We—and 40,000 to 50,000 other families in the Greater New Orleans region—needed to find a place for our children to go to school. We had it easier than most because we had just one child to worry about, and we now knew that our office would be in Baton Rouge. We looked first to that city for a school for Stanford and scheduled a visit for the very next day (September 2) at the Episcopal School of Baton Rouge.

I drove to a nearby Walgreens to purchase a few supplies that we had not brought with us. Our planned three-week motel stay had just been expanded to a disquietingly unknown amount of time. I found an electric pot to boil water for my favorite Lipton tea and inexpensive mugs to drink it in. I also found a Chase Bank close by where I was able to confirm access to our bank account. At least we could get to our money. For the hundredth time in those awful days, I was glad for life's most basic things—a mug of tea and having my sons and husband with me at night! We knew for sure that many people were dead. In time, we would learn just how many. And it was rude to feel anything but grateful.

* * *

Meanwhile, the Millers had basic survival on their minds. The Coast Guard boat brought Harvey, Renee, and Monet five miles away from their safe house to the point where I-10 splits. The trio was dropped off at a ramp, which led to the dry land on the west side of the 17th Street Canal. They walked up to medical stations where people were being registered. There were dozens of trucks parked but no toilets and no food.

Renee, who had had open-heart surgery just one month earlier, was dizzy, so a medical team agreed to take her in. But not Harvey and Monet. Harvey tried to board one of the buses, but no animals were allowed. Harvey walked to the curb, and the obedient Labrador sat down next to him. Someone came by and gave them K-rations (US Army food that heats up by itself). It was red beans and rice, and it tasted horrible. Even the dog wouldn't eat it. Someone else came and gave them a bottle of water. With Monet next to him, Harvey fell asleep on the I-10 entrance ramp with his feet in the road.

* * *

In this surreal world, we got an encouraging dose of reality during the afternoon on Thursday when we received an emailed copy of a forum post from our daughter Aliisa. In the days after the 2005 flood, the forums of the *Times-Picayune* (the local daily newspaper) were, for most, the only mode of communication. The forums, lifelines for much of the citizenry, were posted by NOLA.com, the digital version of New Orleans's single major newspaper. The paper's main office and printing presses on Howard Avenue had flooded, so the paper production was relocated to offices of the *Houma Courier* in Houma, Louisiana, about an hour southwest of the city. Jon Donley, NOLA.com's founder and editor-in-chief, was a visionary who had clear ideas about how the new online medium would be instrumental in putting evacuees in touch with each other and later in rebuilding New Orleans. On this day, the forum post that my daughter Aliisa sent brought us very welcome news: the portion of New Orleans where we lived had not flooded! And it stated that reports of vandalism and looting could not be verified.

* * *

The sound of someone yelling, "Buses are coming!" woke Harvey. He tried yet again to get on a bus, but again authorities said that no dogs were allowed. At one point, a National Guardsman offered to shoot Monet for him—humanely—if that would get him onto a bus.[34] Harvey declined! Meanwhile, twelve ambulances were lined up and waiting, but no one was getting into them.

Harvey walked up to a driver and said, "I will pay you if you take me to Baton Rouge."

"I cannot, sir," he said. "This is the FEDS."

Just then, the driver looked at Harvey's black Labrador and asked, "Is that Monet?" He had grown up in the Lake Vista neighborhood and recognized her. Becoming more helpful, he offered, "Look, we're getting together transportation for the pets because like you, people are refusing to leave without them." The driver then guided Harvey to a kind-looking woman with the Society for the Prevention of Cruelty to Animals.

The woman gave Monet some dog food and then said to Harvey, "I am collecting animals and bringing them to a shelter in Baton Rouge. You cannot come, but I will give you a receipt and later you can get your dog." Then she tried to put Monet into a cage.

The dog had never been caged before, and she didn't bark. She screamed. She screamed and screamed. Harvey took his receipt, tucked it into a pocket, and forced himself to turn his back and walk away as Monet continued to scream. With each step, he told himself that she, at the very least, would be safe.

As Harvey dragged himself away from Monet, a young man, who was driving a bus, pulled up to him and asked, "Are you a single? I have one seat left."

Harvey climbed the steps and collapsed into a seat without even asking where he was going. He learned later that the bus was going to Houston. Harvey didn't know where Renee was, his dog was going to Baton Rouge, and he was now headed for Texas.

* * *

In New Orleans, between 12,000 and 15,000 people had converged on the Superdome and at the New Orleans Ernest N. Morial

Convention Center.[35] Supplies were slowly reaching the Superdome, but none were brought to the convention center. Apparently, Brigadier General Matthew Broderick had access in Washington, DC, to high-tech capabilities but hadn't consulted a basic street map. He thought they were part of the same complex when, in fact, the two structures were a mile apart.[36]

There was another casualty caused by the lack of communication. Rescue of special-needs patients from hospitals and the Superdome was suspended due to rumors of an organized riot. In truth, bands of young black men—many of them armed—were indeed roaming the corridors of both venues, but they were acting as self-deputized sheriffs rather than gangs of marauders.[37] They were searching for food and water. Nonetheless, it contributed to a rumor that there would be an attempt to take control of the buses at ten in the morning. In response, the National Guard blocked busloads of supplies from entering the city because there were not enough soldiers to protect the drivers.[38] When the buses were finally allowed in, it appeared that these same "marauding youths" were the ones who organized the crowds and moved older people to the front of the line.

* * *

At seven o'clock, just as the sun was setting, Harvey's bus arrived at the Astrodome. There were about a thousand cots on the ground floor, all taken. Harvey went to the second floor and saw a woman in a Red Cross uniform. He asked her if he could make a phone call. She replied that along the edge of the stairs were phones and Harvey could call anywhere in the United States for free. In therapy months later, Harvey would recall that the woman looked as though a golden aura surrounded her. He walked to a phone and called his daughter Beth.

"Dad! Where *are* you?" she almost hollered.

"I am in the Astrodome in Houston," he answered.

"I heard from Mom," Beth said. "She's in Lafayette in a hospital."

Beth explained that her husband's parents in Lake Charles were driving to Lafayette to get Renee. Her husband's brother Pat offered to drive to Houston and fetch Harvey. Harvey suggested that Pat try the south entrance because no one seemed to be there. They agreed to meet at 12:30 a.m.

* * *

Thursday night (September 1), Mayor Nagin, who had sheltered with his staff at the Hyatt Regency Hotel, called Garland, who was, as usual, on the air at WWL (AM) radio. It was now more than three and a half days since the levees had broken, and no cavalry was in sight. Nagin's frustration was more than apparent in his tirade which was heard across twelve states.

"This is a national disaster. Get every doggone Greyhound bus line in the country and get their asses moving to New Orleans... This is a major, major, major deal. And I can't emphasize it enough, man. This is crazy... Don't tell me 40,000 people are coming here. They're not here. It's too doggone late. Now get off your asses and do something. And let's fix the biggest goddamn crisis in the history of this country."[39]

The interview was looped for days on the local stations that could broadcast it, on CNN, and on all the networks. The interview commanded attention from leaders all the way to the White House.

* * *

Harvey Miller was still in survival mode. At 12:15 a.m. on Friday (September 2), Harvey sat in the dark at the south entrance to the Astrodome. Down the road, huge searchlights shined. Harvey waited as 12:30 a.m. came and went. Then at 12:50 a.m., Harvey saw a silhouette moving toward him and knew—just knew—that it was Pat.

"He's here to save me!" thought Harvey.

He stumbled to the tall figure, who was indeed his son-in-law's brother, and wrapped his arms around him for a long moment. And, just like that, Harvey changed. Up until then, Harvey's mind was always working, surviving, trying this and trying that. But after Pat found him, it was as though Harvey's mind partially shut down.

For the entire car ride, Pat talked about watching television—or, more accurately, being glued to it—and how no one was able to do anything else. There seemed to be only one broadcast, said Pat, being piped to all the television stations around the nation.

Harvey listened in silence.

* * *

For Steve, Stanford, and me, most of that day (Friday, September 2, four days after the levees broke) was spent in the car. We drove seventy-five minutes from Lafayette to Baton Rouge and checked out the Episcopal High School. But we were too late! The school had already maxed out on the applicants that it could accept. Just then, an unusual thing happened. Stanford's cell phone rang! It was his classmate Reid Chadwick from Isidore Newman School in New Orleans. Reid and his first cousin Mark Allain were staying with family members in New Iberia, twenty-four miles south of Lafayette. Stanford was good friends with both. He and Mark had just enrolled with the Episcopal School of Acadiana, a small town about midway between New Iberia and Lafayette. And the school still had openings for the ninth grade!

We leaped into the car, got back onto I-10, and flew back to Lafayette before turning south onto I-90. Ninety minutes later, we pulled up to the little school. To me, it looked magical with the air of a summer camp. But Stanford was most interested in being in the same place as his New Orleans classmates. After a tour of the school, we met with the principal for about twenty minutes and requested enrollment for the fall semester. The principal was obliging but had a caveat: we needed to pay tuition for the entire year. Otherwise, he said, the school could not make ends meet. We complied. We had little choice.

When the administrative assistant was finished processing Stanford's enrollment, she turned to us and said that she understood how we felt. "I can relate!" she said. She explained that the previous year, while no one was home, lightning struck her house and burned it to the ground. We smiled in appreciation at her attempt to be supportive. But we imagined that she couldn't understand at all. As devastating as that must been—losing her entire home and its contents—at least she still had her family, friends, neighborhood, and all the things that create a community. We said goodbye, climbed into the car, and drove the thirty minutes back to the Drury Inn. As always, we turned on the radio and listened to the nonstop coverage of the levee failures and flooding.

* * *

Garland raged at the indignities. What was taking the cavalry so long? Two days after Hurricane Charley hit Florida in 2004, FEMA had moved two million meals into Florida and 8.1 million pounds of ice.[40] It was now *four days* after the hurricane's winds had died. A Republican-dominated House investigative committee would later note, "We cannot ignore the disparities between the lavish treatment by FEMA of Hurricane Charley survivors and the survivors of this 2005 hurricane."[41]

Later that day (Friday September 2), one thousand troops finally arrived. Their commander was Lieutenant General Russel Honoré, a cigar-chomping, no-nonsense leader.[42] He brought the army's 82nd Airborne and 1st Cavalry Divisions. After stopping at the Superdome, he walked the half block to the staging area for the troops and talked to guard commanders. These troops were not under his command and were expecting to see chaos. But he told them to keep their weapons lowered at all times.

"This is not a military operation," the lieutenant general barked. "This is not Iraq."[43]

Video footage shows Honoré ordering them to lower their guns and cursing at those slow to obey. Lieutenant General Honoré's strength was his restraint.[44]

With the ever-present cigar, Honoré got into the lead vehicle, and the convoy headed toward the convention center at 12:25 p.m. and was greeted with cheers. One little boy saluted. It took about twenty minutes for the general to establish control and set up six food-distribution stations along the length of the convention center. A team of Arkansas guardsmen were assigned to sweep the enormous building. A hundred people were splayed out on the floor, nearly dead of heat, dehydration, and starvation. But they found no gangs of thugs.

As described by Jed Horne with the *Times-Picayune*,[45] "Rumors of gang rapes and wanton murder needed to be repeated only two or three times before reporters decided the rumors had been corroborated and repeated them in print." In the end, four bodies were found, all died from natural causes.

Finally, President George W. Bush was going to make an appearance in Jackson Square. A gesture like that meant a lot to all of us—to those who evacuated and those who stayed.

* * *

Away in Texas, Harvey Miller had, at last, reached safety. At the house, Pat showed Harvey to a guest room with a connecting bath. Harvey refused food but took a hot shower. He went to bed and fell immediately to sleep. He woke up an hour later, got up, and took another shower. A few hours later, he took another shower. Months later, in therapy, Harvey said that he felt like he needed to wash everything away. Harvey slept until almost noon on Friday (September 2). When he awoke, he got the word that Renee was in Lake Charles, safe and sound.

* * *

With Stanford now registered in a school, we returned to the Drury Inn. As we walked down the long hotel hallway, we could hear our little dog barking. Reunited with Chester and our older son Mark, we returned to the obsessive-compulsive activity of watching television.

By now, Wal-Mart was in New Orleans,[46] and supply trucks were making regular runs to the Louisiana Superdome and newspaper reporters and television news crews were moving around the city.[47] But FEMA's Red October still had not arrived, sitting now in Baton Rouge eighty miles away.

In contrast, our world, while turned upside down, was quite comfortable in a Lafayette hotel. Our biggest challenge was competing for its single coin-operated washer and dryer. In other words, our troubles were small and petty compared to the majority of the Greater New Orleans's survivors.

Dinner that night was an excellent meal, stewed turkey and gravy over rice, courtesy of the Red Cross. Over our comfort food, we discussed how we had succeeded in figuring out two of life's most basic things: where Steve and I were going to work and where Stanford would go to school. It was now time to figure out where we would live. Indeed, we should have been working on this much sooner, but that was impossible without knowing where Stanford would go to school. Unfortunately, it was already too late. Everything rentable was snatched up. Our only option was to purchase a house or condominium. And even then, we were behind the eight ball.

* * *

Saturday morning (September 3), after breakfast in the crowded hotel restaurant, I was startled by the sound of my cell phone ringing. It had become a "texting machine" and not a phone. I answered and heard the voice of my youngest brother Mike, a firefighter in North Attleboro, Massachusetts. This contact was the first I had with any of my family members since the levees broke.

I quickly explained that my own family and my husband's relatives were all safe and that none of our homes had flooded. Then I asked my brother if he had been following the news reports about the flooding in New Orleans.

"Sandy, you've got to be kidding!" Mike almost shouted. "That's all we do! That's all everyone is doing!"

Unbeknownst to us, practically every person in the nation had been glued to the television since Monday (August 29). This was especially the case now, five days after the levees broke, because it was the Labor Day holiday weekend. I told Mike that I was glad that this was getting a lot of national attention. With the community-oriented attitude that seems universal in firefighters, he wanted to come to New Orleans and help with the rescue. I explained that the feds were now present in the city and that everyone was being evacuated. I believed that the best thing he could do was assist the evacuees in his region.

"It's a terrible disaster, Mike. Everyone is in shock. People will be talking about this flood ten years from now."

I had no idea that the resulting 2005 flood would likely be talked about for the next hundred years, no different than the sinking of the Titanic in 1912 which took the lives of 1,517 people, roughly the same as the 2005 disaster.[48] And, like the Titanic, the levee-breach event was a pivotal moment in history. Directly due to it, changes to US law were passed that improved life for the 55 percent of the American population living in counties protected by levees.[49] But, of course, I knew nothing of these things yet; I was still in survival mode.

All across the City of New Orleans, floodwater was penetrating more than homes and buildings; it was invading electrical wires, fiber-optic cables, and natural gas lines. Nearly all the city's pumping stations were damaged. Hospitals lost emergency power when generators and fuel tanks flooded in basements. City and state building codes required generators, but codes made no specific mention that the generators should be located above the floodplain, as should electrical switching equipment and fuel.[50] About 134,000 housing units were severely damaged in New Orleans alone.[51]

* * *

Harvey Miller finally learned that Monet was safe at a kennel in Baton Rouge. Another one of Pat's brothers drove there and fetched her. On Saturday morning (September 3), Harvey was reunited with Renee and a very happy Monet in Lake Charles. The following Monday (Labor Day), the Millers relocated to Little Rock, Arkansas, because they had a daughter and son-in-law there. The Red Cross offered to put them up in a motel for two weeks.

* * *

All day Saturday (September 3) and Sunday (September 4), we drove around the city, looking at impossibly expensive houses to live in until we were allowed back home. They were all much too large. We were too late in starting the hunt. However, at 7:15 a.m. on Monday morning (September 5), a lucky break came our way. Steve had been watching real estate notices online, and on Labor Day there was a new listing—by owner—at a less outrageous price. We ran to the car and sped off to a house on North Roclay, in a subdivision west of Lafayette. At precisely 8:00 a.m., we rang the doorbell and the owner—a young woman with a cheerful, bright face—answered. Her name was Mrs. Thibodeaux, and she showed us around the home. The asking price was more than we wanted to spend, but we signed the paperwork on the spot. Just as we were signing, another couple pulled into the driveway, also intending to look at the house.

Later, Ms. Thibodeaux would confide that the couple, upon hearing that we had just signed a contract, had said to her, "Whatever price you agreed on, I will pay you $10,000 more."

This illustrates how desperate people had become while trying to put a roof over their heads. Ms. Thibodeaux could have taken the larger offer, but she didn't feel that it was the right thing to do.

We now had found solutions to our three basic questions: Where would Stanford go to school? Where would we work? Where would we live? All were decided upon at breakneck speed. But we were lucky in our quickness to act. Many people were so shell-shocked that even thinking was just too much to bear, never mind *acting*.

* * *

The Monday after the 17th Street Canal levee broke (September 5), the breach was sealed with seven thousand sandbags, each weighing ten thousand pounds.[52] Dewatering the city was originally expected to take months, but the weather remained hot and dry and a significant fraction of the water evaporated.[53] Even so, the Army Corps reported pumping out 250 billion gallons of water.[54] Meanwhile, businesses in New Orleans were scrambling to find employees. Burger King was offering six thousand dollar signing bonuses to anyone who agreed to work for a year at one of its New Orleans outlets.[55]

Had the levees not broken, my older son Mark would have returned to college the following Thursday (September 8). However, Mark decided to return to Denver early and stay with his grandmother until the university opened for the 2005–2006 academic year. He was not enjoying life at the Drury Inn among hundreds of traumatized people. The airline allowed us to book a ticket out of Ryan Field in Baton Rouge to Denver without a penalty.

On Monday afternoon (September 5), I drove the one-hour trip to drop off Mark. I offered to stay with him until the flight, but he insisted that he was fine and I shouldn't wait with him. As I hugged him goodbye, I sensed that Mark wanted—as soon as possible—to put the whole traumatic mess behind him.

We all wanted to. Few could. Some never would.

* * *

The Millers were allowed to use the motel's breakfast bar every morning, and local church groups brought dinner for them each night. They were safe and surrounded by caring people. But Harvey was not himself. He spoke little and allowed himself to be led around.

* * *

The next day was Tuesday (September 6, the day after Labor Day), the maiden day for the Rosenthal family to practice one of the weirder post-flood rituals: waking up in a hotel room and preparing to send a child off to school. One does not normally awake in a hotel room before feeding a child his breakfast, fixing his sandwich for his bagged lunch, and driving him to the school bus stop. Even so, the ritual of waking and preparing to send a child to school provided another touchstone to reality. It felt normal and familiar and helped keep us sane and focused during those insane days.

We discovered how little we needed. Some basics one cannot live without. For example, I had to have boiling water and a mug for my Lipton tea in the morning. I had to have my laptop. And in that turbulent time, a television set became necessary. But, beyond that, as long as my husband, my son, and my little dog were nearby, I was content. A year later, upon reconnecting with friends, we would all report this same feeling. All of us would report feeling a need to "purge" ourselves of unneeded things. Even people whose houses had flooded and who had lost everything felt the same way.

On Wednesday morning (September 7), Dan Silverman, one of the senior members at Steve's business, and Dave LaBruyere, the company comptroller, drove back to the flooded New Orleans region to check out the status of the Metairie office. They found that the building had taken on a couple of inches of water. Under normal circumstances, an inch or two of water would be considered "inconvenient" because the region's pumping system could have removed the water in a few hours. But in this case, the water would sit for weeks because the pump-station operators had been sent away.

In the days leading up to the 2005 flood, the president of Jefferson Parish chose to adhere to its now defunct "Doomsday Plan," which evacuated two hundred drainage-pump operators the day before the hurricane made landfall.[56] Without a pump system, the water sat and allowed mold to accumulate. With closed windows and no ventilation, mold growth became a significant challenge that required extensive cleaning and disinfecting. In addition, the mold reached levels associated with respiratory symptoms and skin rashes.

After finishing their assessment of the office, Dave and Dan drove to our house and, using the key we gave them, let themselves in. First, they checked for flooding. As we had hoped, the flooding did not reach our house, stopping about six blocks away. Then, they emptied the refrigerator and freezer.

Our placing blocks of ice in the freezer and refrigerator to buy time would have worked had we been gone for just a couple of days, but a week was too long. The food was rotted, and all had to be discarded. But there was an unexpected benefit. The ice blocks saved the appliances from being destroyed by rotting food. Nearly everyone else in the city had to throw out their large appliances. But the ice blocks saved ours.

The dynamic duo then proceeded to their own homes before returning to Lafayette. Now, we could stop thinking about the status of our house and start focusing on getting through the next three months. We had just learned that our son's school in New Orleans—Isidore Newman School—had announced that it would reopen in January 2006. So we set our sights on returning home after Stanford completed his fall semester at the little life-saver Episcopal school.

* * *

On Saturday morning (September 10), Stanford, Steve, Chester, and I went on our own trip back to New Orleans. There was the natural, albeit morbid, curiosity to see it for ourselves. And we also needed to get my car, which we had left to help relieve congestion on the interstate during the evacuation. We needed to check on the fish, and Steve wanted to get some pots and pans for cooking when we moved into the house.

The mood was cheerful enough as we traveled through the westernmost suburb of New Orleans to the city of Kenner. The famous, blue-tarped FEMA roofs were not yet installed and, to the untrained eye, the damage in Kenner and Metairie did not look serious. But our mood changed as we entered New Orleans. The great outdoors went from green to gray in a second. We had passed from the west side of the 17th Street Canal to the east side.

Driving was tricky because no traffic lights were working and downed trees blocked most roads and streets. We zigzagged our way down St. Charles Avenue to our home on Soniat Street. The garage door was down and, since there was no electricity, we entered the house through the front door. Search and rescue teams had marked the door with the omnipresent "X" and code numbers to indicate whether any people—or bodies—were inside.

The first thing we noticed in the front garden was that the wind strength wasn't enough to knock the papaya fruit off our trees—a testament to the relative weakness of the hurricane's wind in New Orleans. We opened the door. Even though the September day was blazing hot, the house was rather cool due to the boarded-up windows. Chester ran to his water bowl, and Stanford ran upstairs to check on the fish, who were all accounted for. Stanford replaced the batteries in the tank's aerator and gathered up the clothes and items he wanted to bring back to Lafayette. At the other end of the house, I packed up my small desktop computer and several framed family photos.

Steve manually opened the garage door, I backed my car out of the driveway, and Chester hopped back and forth from my lap to the passenger seat. We picked our way to the house of our friends, Cherry and James Baker, who lived at the corner of Napoleon and Claiborne Avenues. Their home was elevated six feet from ground level, but still they had flooded. Two large motorboats rested on the dead, gray grass, moored to the porch rail in front of their house. Steve made a futile attempt to call James on his cell phone. Amazingly, he got through on the first try. "I have good news and bad news. The bad news is that you had a foot and a half of water in your house."

"What's the good news?" James asked.

"You now own two motorboats!"

This is how we survived. If you are going to hear the news from one of your closest friends about how much water your house had sustained—something that houses are not built to handle—you needed to hear the news with a joke. Humor relieved the nagging tension that clung to your very bones, to your cells. So many unknowns, so many unanswered questions, so many whys! And, most nagging of all, what was going to happen to all of us? Gentle humor made it bearable.

* * *

We got back into our two cars and continued our strange trip. We wanted to see the site of a levee breach. We tried to access the 17th Street Canal, which was about three blocks from one of our favorite tennis courts. To get there, we took the I-10 because it was the only dependable roadway.

As we exited at West End Boulevard, we looked up in shock to see a gigantic pile of debris! We continued down the boulevard toward the lake. But, as soon as we turned onto Old Hammond Highway, we were intercepted by armed guards, who told us that access to the 17th Street Canal was blocked because the Army Corps was doing emergency repairs to the levee and floodwall. It was for our own safety, they said. We would learn later that the Army Corps also blocked independent levee investigators, who flew in from California with the very same logic.

Having seen enough lifeless houses, downed trees and telephone poles, Xs on doors, and the sad, dead color of gray, we decided to return to our Land of Oz.

As we rode back to Lafayette, a "secret meeting" was taking place at the Anatole Loews Hotel, just north of Dallas. The room was reserved by James (Jimmy) Reissa, a wealthy, uptown blue blood who was chair of Mayor Nagin's Regional Transit Authority. The meeting was attended by Nagin and members of the Business Council of New Orleans: Dan Packer, CEO (chief executive officer) of Entergy, and Joe Canizaro, a real estate developer and personal friend of President George W. Bush. Others from the Business Council included Jay Lapeyre, owner of Laitram Industry and the Greater New Orleans Business Council chair, Gary Rusovich, Gary and George Solomon, and William Boatner Riley.

During the Dallas meeting (September 10, which was twelve days after the 2005 flood), Canizaro had Karl Rove—senior advisor and assistant to President George W. Bush—on the phone. Rove had a directive for the mayor: "Put together a blue-ribbon panel of businesspeople and community leaders who would vouch for a rebuilding before the federal government committed to spending tens of billions."[57]

In later interviews, Nagin's attorney would describe the ensuing conversation as "a lot of dry talk of the best way to dewater the city, fix the infrastructure, and rebuild the battered business community." Nagin, in his memoir, would recall that the businessmen in the suite reserved by Reiss were "intent on engineering a very different New Orleans."[58]

* * *

The new temporary office in Baton Rouge opened on Wednesday (September 14), due to a gargantuan amount of work on the part of Leslie, Steve, and many others who were working for Strategic Comp, and due to the remarkable generosity of Great American.

* * *

Help and generosity were extended to the Millers as well. Harvey and Renee accepted the Red Cross's largesse of putting them up in a hotel in Little Rock for two weeks. During that time, they, with the help of their daughter Beth, found an apartment. They moved in on September 19. Despite her recent surgery, Renee fared well, but Beth was worried about Harvey. She noticed that her father wasn't talking and would answer questions only in monosyllables. When Harvey didn't improve, she became frightened. Beth called a psychiatrist friend who suggested that Harvey see a trauma counselor. Harvey might be in shock.

* * *

As soon as the Thibodeaux family moved out of the North Roclay home, we could move in. The target date was September 30 because the new house they had bought would not be available until then. Of course we were dismayed, but the housing shortage created this weirdness. For the next twenty days, we continued the odd life in a hotel. Stanford went to school during the day, and Steve spent all day keeping his workers' compensation business operating from a laptop. At night, I tried to create a normal environment for Stanford to do his homework in our hotel room by rearranging the furniture and placing framed photos under the lamps.

* * *

As we and 100,000 other families struggled, a horrified world was demanding answers. From the start, the Army Corps had told CNN and other big media outlets that the storm surge during the August 2005 hurricane was greater than what the floodwalls were designed to withstand.[59] But on September 21, 2005, Michael Grunwald with the *Washington Post* cited computer models and eyes-on-the-ground evidence from scientists and engineers at LSU Hurricane Center. They concluded that storm surge had not come even close to flowing over the tops of the floodwalls along the 17th Street and London Avenue Canals.[60]

Former senator J. Bennett Johnston (D-LA), who was influential in pushing the Army Corps to build floodwalls along the London Avenue Canal,[61] was surprised: "It shouldn't have broken."[62]

When Johnston provided his comment to Grunwald, he recalled numerous briefings from Army Corps officials about the danger of a hurricane overtopping the New Orleans levees. But he said that he never envisioned wholesale breaching: "This came as a surprise."[63]

The implications of this report were dire for the Army Corps. It could mean that improper design caused the failures of these levees. Right after Grunwald's above-the-fold story in the *Washington Post*, Governor Blanco arranged a meeting with LSU Hurricane Center experts and with Johnny Bradberry, Secretary of the Louisiana Department of Transportation and Development. The group selected a tousle-haired professor from South Africa, deputy director of LSU Hurricane Center Ivor van Heerden, to lead an investigation of the levee breaks. It was called Team Louisiana, and van Heerden was provided a budget of $130,000.[64] The McKnight Foundation also provided $250,000.[65]

Dr. van Heerden might be the first person to have stated publicly that the Army Corps was not being honest with the public. The outspoken deputy director claimed that, for New Orleans, the 2005 storm was a "decidedly mild" hurricane. "Call it a blame game if you must," Dr. van Heerden said, "but some of us were determined to find out exactly what happened and to demand justice from the responsible parties."[66]

Van Heerden did not support the Army Corps' statements to the public that the hurricane had been a monster storm that was just too huge to hold back. But his early statements were educated guesses, not investigative conclusions. At this early stage, it was impossible, even for the experts, to wrap their heads around so large a disaster.

* * *

I drank up these news stories at night on my desktop while Steve worked on his laptop, holding his company together, and Stanford studied. I was reading, reading, reading about the 2005 flood. This odd ritual went on for a month. Every day was pretty much the same except for one memorable interruption: Hurricane Rita.

The September storm formed in the Gulf of Mexico only a few weeks after the August storm, making it the first time on record that two hurricanes reached Category 5 status in the Gulf of Mexico in one season. Initially, it was headed toward west Louisiana where we—and thousands of others—had evacuated to. Grandma Rose was moved to safety several days before the storm arrived. But at the Drury Inn, the mood was calm among the evacuees, due perhaps to the realization that, in New Orleans, the most harm done was due to water that flowed through the breached levees. At the Drury Inn, we had the protection of steel-reinforced concrete far from water or levees.

On Wednesday (September 21) in the neighboring State of Texas, the mayor of Houston made an official call to residents to flee, embellishing his pleas by adding, "Don't follow the example of New Orleans and think someone's going to come get you."[67] But Houston experienced an evacuation that was rife with gridlock, emergencies, and empty gas stations. A subsequent *Houston Chronicle* article reported that people would have been better off staying home. The evacuation contributed directly to sixty deaths, twenty-four of which occurred when a bus, carrying nursing home residents, caught fire and exploded on I-45 near Dallas.[68]

Stanford's school was closed the day before Rita's arrival to allow for preparations. We took advantage of the day off to attend to an important milestone in a young man's life: getting his learner's permit to drive a car. During the paper shuffling and processing at the

Department of Motor Vehicles, all the talk around us was about the impending storm. NOAA had just declared it one of the strongest storms on record for the Atlantic Basin, with peak sustained winds at 175 miles per hour. On the night of September 24, 2005, Rita made landfall on the Texas-Louisiana border as a Category 3 storm with wind speeds of 120 miles per hour.

Hurricane Rita's worst damage was the southwestern area of Louisiana where storm surge topped agricultural levees and flooded low-lying coastal parishes.[69] Rita drowned thousands of cows and dispersed others ten miles inland or more with only brackish water to drink. In New Orleans, the Industrial Canal breached again and flooded the Lower Ninth Ward for the second time in a month.

Our next day in Lafayette was unlike our bewildering days after the New Orleans flood. It was just a day of heavy rain with nothing to do because businesses and restaurants were ordered closed.

* * *

In Washington, DC, on September 28, the House Appropriations Subcommittee on Energy and Water Development hosted a hearing associated with the August 2005 flood.[70] Members of Congress—who were responsible for how much aid the people of Greater New Orleans would receive—needed accurate information. This may well have been the single most important hearing for the people of New Orleans' welfare.

Anu Mittal, Director of Natural Resources and Environment, testified on why the levees were still not complete when storm surge arrived. She read from a script, which she later destroyed and which bore little resemblance to the General Accounting Office

(GAO) report she had submitted that day. Here is a transcript of a key excerpt:

> After the (levee) project was authorized in 1965, the Corps started building the barrier plan…parts of the project faced significant opposition from local sponsors and *they did not provide the rights of way that the corps needed* to build the project on schedule. But most importantly there were serious concerns relating to environmental impacts of the control barriers *that were supposed to be constructed*…on the tidal passages to the lake. This ultimately resulted in a legal challenge and in 1977 the courts enjoined the corps from constructing the barrier complexes. [Emphasis added.][71]

The effect of the testimony was lethal. The clear takeaway from Ms. Mittal's verbal testimony was that, in the years leading up to the levee breaks, the Army Corps struggled to complete the flood protection on schedule and were hamstrung by local officials and environmentalists. In fact, Ms. Mittal's accompanying GAO report stated none of these things. But the damage was done. In the videotaped recording, the listeners sitting behind Ms. Mittal can be seen registering expressions first of being appalled and then of indignation.[72]

Members of Congress at this hearing were led to believe that the people of New Orleans had brought this levee-breach situation upon themselves. One Congress member stated, "It's obvious. It's their fault!"[73]

Where did Ms. Mittal's verbal testimony come from? Years later, after painstaking research, I figured out that the material Ms. Mittal presented in her verbal remarks was loosely based on what was, at that time, a twenty-two-year-old GAO report.[74] Ms. Mittal had presented cherry-picked data from statements made by Army Corps officials in 1982. Ironically, the document's final judgment, even with its tone of neutral "government speak," reprimanded the Army Corps officials for their glacial pace in building the levee-protection system for New Orleans in light of the critical need for project completion.[75]

In recorded audio ten years later, Ms. Mittal readily admitted that her verbal testimony in September 2005 was "from three or four decades ago."[76] When asked if she still had a copy of her verbal testimony from September 28, 2005, she said no. When asked if she remembered who assisted her in preparing her verbal testimony, she said it was not retained. In my mind, someone had coached Ms. Mittal. The Army Corps has an established and documented policy of creating talking points for influencing people in high places.[77]

The damage from Ms. Mittal's verbal testimony cannot be understated at a time when all eyes were watching and all ears were listening. In a phone interview on November 10, 2015, Sen. Mary Landrieu (D-LA) confirmed that videos of post-disaster congressional hearings—like Ms. Mittal's—were circulated and became highly influential on members of Congress. When queried on whether she believed that this testimony may have biased or prejudiced Congress members against the people of New Orleans, the senator responded, "It was well understood that there was pressure from the White House, constantly and early on, to assign blame on the residents of New Orleans for the failure of the levee system."

So, it began. Four weeks after the flooding that took the lives of at least 1,577 people[78] and nearly drowned an entire city, our Congress had a false perception. And every resident of New Orleans would soon pay the price. The local officials were considered guilty until proven innocent. Years later, I searched the internet and could locate no congressional record of Ms. Mittal's testimony. I also learned that no transcript of the hearing was sent to the Government Publishing Office. The hearing was also missing from a list compiled by the American Geosciences Institute.[79] If it were not for C-SPAN—a private company—all information about the first congressional hearing may have been gone forever.

* * *

The next day was a turning point for me. My husband forwarded a written story about Anu Mittal's testimony to me that contained a link to download the full GAO report. Ms. Mittal's written GAO report explained the history of the levees that encircled New Orleans. The key factor was that the Army Corps was charged by Congress for the design and construction of the levees. And the local sponsor was responsible for maintaining the structures once completed. Period. There it was in black and white. Energized by this report and by Ivor van Heerden's admonitions in Grunwald's *Washington Post* article, I formulated my own version of the flooding event: the federal government was responsible.

Two days later (September 30, one month and one day after the 2005 flood), Mayor Nagin unveiled his Bring New Orleans Back Commission (BNOBC) at a press conference in the high-rise Sheraton Hotel on Canal Street. Cochaired by Maurice (Mel) Lagarde (a white blue blood) and Barbara Major (a black woman who was raised in the Lower Ninth Ward), the commission was composed of fifteen men, seven of whom were CEOs. Three were bank presidents.[80] Jimmy Reiss, chairman of the New Orleans Regional Transit Authority, had recommended that the commission also include education leaders, so Scott Cowen, president of Tulane University, was added.

* * *

Back in New Orleans at about this time, the Army Corps sent in teams of engineers to collect data in order to explain the levee breaches while it was also doing emergency breach-repair. The repairs were causing crucial data to be degraded. As this was happening, crews of independent civil-engineering experts attempted to access the levee-breach sites. But Dr. Paul Mlakar, senior research scientist with the Army Corps' Engineer Research and Development Center (ERDC), denied entry to all of them over concerns for their "safety" and concern that their presence could impede emergency operations.[81]

One of these outside crews denied entry was an elite team assembled by the University of California, Berkeley, and funded by the National Science Foundation (NSF). It was called the Independent Levee Investigation Team (the Berkeley team),[82] chaired by civil-engineering experts Raymond Seed and Robert Bea with the University of California as well as Dr. J. David Rogers, professor at the Missouri University of Science and Technology.[83] The trio found it strange that Dr. Mlakar repelled them. After all, they were specialists with extensive experience in early arrival at major catastrophes, including earthquakes in developing countries where there were thousands of bodies in rubble and where issues like sanitation and safe water were even bigger problems than in New Orleans. Often, there were frenzied emergency-rescue attempts still going on. After all, these experts understood field operations and how to stay out of the way. Besides, they were a tiny group.[84]

Chapter

3

The Fairy Tale

October 1, 2005, was a day of celebration! My husband, my son, and I moved into a little one-story house at the westernmost edge of Lafayette. At the grocery store, I bought ingredients for our first glorious, home-cooked meal in what seemed an eternity. While my husband prepared dinner, we realized we didn't have an important condiment: salt. I ran to the house next door to introduce myself and borrow some. An hour later, Stanford, Steve, and I sat down in a real kitchen, like normal people do, and savored the most delicious chicken and garlic à la Mosca with rice. It was so good to be out of a hotel room! But we were aware of how lucky we were, and we continued to reach out to friends and invite them to stay with us if they were having trouble locating a place to live until they could return home.

Since we had a high-energy dog who needed multiple daily walks and since we were gregarious New Orleanians, we managed to meet every neighbor within a half mile. And when Steve made some of his famous homemade pralines and gave them away as gifts, that sealed the deal. We all became friends.

This was how we met Mr. Bouillon, our neighbor and savior for Stanford. The first question he asked my son was what sport he played in school. (Mr. Bouillon was the local high school's physical therapist.) Stanford explained that he had not played any sports for two years due to a case of shin splints so severe that even walking was painful. Excused from all sports and physical education classes, Stanford had made the best of the extra time by teaching himself computer skills.

Mr. Bouillon listened and said, "But you're missing your childhood! You should be running and jumping!"

Then Mr. Bouillon asked me if he could examine my son. For a few seconds, I recalled the three orthopedic physicians whom I had already brought Stanford to see. And then I responded, "Of course, you may!"

And off they went to another room. Fifteen minutes later they returned, and Mr. Bouillon asked me if he could take Stanford's orthotic inserts back to his home and adjust them. After all the doctors and physical therapists we had seen already, I replied that if he thought waving a chicken bone over Stanford's head would help, I would say yes.

Mr. Bouillon smiled in apparent understanding and took the inserts to his home. He returned an hour later and gave them almost ceremoniously to Stanford. In my mind's eye, I saw a cloaked wizard handing Stanford a pair of magical shoes.

"Wear your inserts to school tomorrow," he said. "When you get home, try running. Just try it and see what happens."

The next day went like all other days. We got up at 6:05 a.m. I fixed Stanford his favorite sandwich for lunch—smoked turkey on a toasted roll—and Steve drove him to the bus stop. I read and read and read all day. Promptly at 3:45 p.m., I picked up Stanford at the bus stop.

When we got home, he said, "Well, Mom, I'm going to go outside and try running."

I smiled and said, "Okay!" I stayed inside, so he wouldn't see my look of disappointment. I did not expect anything to be any different.

A moment later, Stanford burst back into the house, shouting, "Mom! I can run! And it doesn't hurt!"

I leaped up, and together, we ran out into the bright sunshine. And Stanford ran! And then he laughed! And then ran some more! And laughed some more! And then Stanford, who at that time was a shy fifteen-year-old, ran up to me and hugged me!

* * *

Later that afternoon, after sharing the amazing good news with every family member I could reach with our still-intermittent phone service, I bought some stationery and a small box of chocolates. I suggested that Stanford write a note to Mr. Bouillon, and then I would deliver both. When Stanford brought the note to me, I asked him if I could read it. He nodded. Here's what it said:

> Dear Mr. Bouillon,
>
> I did exactly what you told me to do. I wore my inserts to school, and then when I got home I tried running. I could run, and it didn't hurt! I have been dreaming about this day.
>
> Thank you very much,
> *Stanford*

If you talk to any survivor of the 2005 flood, most will share a tale of a silver lining, a way in which the survivor's life was improved in some way. For me and my husband, our lives were made better because the 2005 flood put us on a straight-and-narrow path to Mr. Bouillon's doorstep. Our son was no longer in constant pain, and the world of sports would soon open back up for him.

* * *

The Millers also needed medical care. Their daughter Beth continued to worry and urged both of them—especially Harvey—to see a trauma counselor.

* * *

Away in Washington, DC, the American Society of Civil Engineers (ASCE), an elite engineering trade group, was working out an arrangement with the Army Corps. Larry Roth, the ASCE's executive deputy director, was doing most of the deal-making. The agreed-upon arrangement was that the ASCE staff would handpick a group of experts to perform an external peer review for an Army Corps-sponsored levee investigation, which was yet to be announced.[85] This arrangement would guarantee the trade group a "place at the table" and a two million dollar fee from the Army Corps.[86] Worded differently, the Army Corps was making its self-investigation appear to be "independent" yet all the while had chosen its peer-review team and was paying them, too. Both were a conflict of interest. And no one knew anything about it yet.

* * *

On the other side of Washington, DC, firefighters from New Orleans testified about what they saw on August 29, 2005. Captain Joe Fincher and rookie Gabe King had witnessed floodwaters flowing through the breach of the 17th Street Canal.[87] They had firsthand knowledge and Captain Fincher had videotape.[88] As soon as the winds died down enough, they had swum out to find a boat to commandeer. Captain Fincher had hot-wired it, and they made life-saving trips into the neighborhood near the breach. The rescues continued for four days, from sunup until dark. Their testimony was riveting, but they were ordered to be quiet until the Army Corps' investigation was completed.[89]

* * *

In New Orleans, after being rebuffed multiple times, the Berkeley team—led by Drs. Seed, Rogers, and Bea—was finally allowed into the field. But there was a caveat: they had to be escorted by a team from the Army Corps led by Dr. Mlakar. According to Dr. Seed, Mlakar's role seemed to be to "keep the Army Corps personnel from speaking too openly with the rest of us and thus potentially spilling any beans."[90]

Almost three hundred miles of levees had to be examined. The urgency on the part of the Berkeley team was intense because data continued to disappear daily. In response, the now-formed ASCE team announced that they would cooperate with the Berkeley team to prevent redundant work and maximize efficiency. This move made sense since most of the experts had known each other for years.

The Berkeley team cochairs and the ASCE team members prepared a semi-formal outbriefing for the Army Corps press conference, which was scheduled at the 17th Street Canal. A problem arose, however, because it seems that, because of a deal between the ASCE and the Army Corps, the field teams were not permitted to discuss with the media what they had learned. They could only discuss what they had "measured." Since the Berkeley team had already agreed to work with the ASCE team, they were muzzled as well.

As expressed by Dr. Seed in his forty-two-page ethics complaint filed in 2007, "It was ethically and professionally offensive to the two assembled teams of experienced experts to be told that they were to simply wave, say that they had measured things, and that they had learned nothing. And at a time when a distraught population and the government (both local and federal) were in desperate need of some small sense that engineers were performing a straightforward, honest investigation and were making some progress."[91]

* * *

Six weeks after the 2005 flood (October 11), the City of New Orleans was officially dewatered.[92]

Seven weeks after the 2005 flood, despite multiple investigating teams, the surface had yet to be scratched on the who, what, where, and why of the levee-breach event. If one were to count the breaches on a graphic map created by the Army Corps, they would find a total of fifty-two breaches in the region.[93] It was impossible at this point for any human being or group of human beings to draw conclusions on what happened in so complex a scenario. Communication lines were still down, and breaches needed to be plugged.

Yet, just seven weeks after the levees broke, the Business Council of New Orleans had decided where the fault lay. On the day the floodwalls broke, this small group had no phone number, no staff, no meeting minutes, no list of expenses, and no membership list. But, with the city barely dewatered, they had already decided that blame belonged to the Orleans Levee Board—people whose chief responsibility regarding floodwalls and levees was maintaining them after the Army Corps built them. By October 20, while some souls were yet to be discovered in their attics, this group had already submitted a bill to the Louisiana legislature. It was sponsored by State Senator Walter Boasso in St. Bernard Parish and it would change the way members of the Orleans Levee Board were selected.[94]

Up until 2005, all Levee Board members were selected by the governor. Boasso's bill would take control of who selected the

Orleans Levee Board members out of the governor's hands and give it to a small group of people with life terms, a so-called "blue-ribbon committee" that would select its own people.

Seven weeks is far too soon for anyone to wrap their heads around what had happened, let alone figure out a way to fix the problem. For example, the investigation of the I-35 bridge collapse over the Mississippi River in Minneapolis took eighteen months in a situation where the bridge was completely closed down. The Rogers Commission Report on the Space Shuttle Challenger accident took more than four months to complete in the face of intense national pressure.

But here, just seven weeks after fifty-two levee breaches flooded forty-eight square miles of a metropolis, a bill was crafted and ready to go without one single completed levee investigation. It would appear that plans to change the way the Orleans Levee Board was selected were discussed long before the 2005 flood. As noted by Naomi Klein in her book, *The Shock Doctrine: The Rise of Disaster Capitalism,* changes to policy are often pushed through while citizens are in shock from disasters, upheavals, or invasion.[95]

The business community was loud in its condemnation of the Orleans Levee Board but named no names. The many dozens of men and women who had served for decades on the board had no faces. Looking at the list of eighty-four men and women who had served on the Orleans Levee Board since 1925, one would see all kinds of upstanding people: city mayors, city councilpersons, and philanthropists. One Levee Board member who served from 1997 to 2001 was a Catholic nun: Sister Kathleen Cain, a provincial with the Franciscan Missionaries of Our Lady.[96] But the Business Council of New Orleans had lumped them together as one giant cohort of corruption.

* * *

The media, aware of what stories get the most eager readers, began to focus on Jim Huey, who was acting president of the Orleans Levee Board on the day the levees broke. The opening line of a *Los Angeles Times* story on October 28, 2005 hollered: "The president of the Orleans Levee Board, who played a key role in decisions about the construction of levees that failed during Hurricane Katrina, resigned Thursday."[97] The article criticized a no-bid contract that Huey gave three days after the levees broke. Huey had leased three thousand square feet of office space in Baton Rouge from his wife's cousin. His explanation was that the board's lakefront headquarters were damaged by storm surge and the state government had failed to provide an operation base. Eyebrows were further raised when it was revealed that, three weeks before the flooding, Huey had applied for back pay that he was owed for nine years of service. All Levee Board presidents, including the post-storm president, receive a stipend, but the news report, which went viral nationally, claimed wrongly that he took his earned back pay "illegally."[98]

Years later, Huey and I had coffee to discuss the accusations. Together, we recalled the Wild West days after the flood when finding office space was paramount. Quite possibly, immediately leasing space from his wife's cousin saved the state a lot of money as space was being snatched up at lightning speed. Huey also told me that his big mistake was returning the back pay to the state. Had he instead turned it over to a lawyer and undergone legal scrutiny, he would have been vindicated. But the damage was done. Huey told me that he resigned on October 27 because he didn't want the issues, which were irrelevant in the flooding, to become a "circus sideshow." In his exit, he defended both the lease and the back pay, and welcomed investigations into their legality. No wrongdoing was ever found.

More important than the claims about the nepotism and the back pay is this: Jim Huey's role as president of the Orleans Levee

Board was overseeing maintenance. Only the Army Corps made decisions about levee design and construction.

* * *

In October 2005, "that corrupt Orleans Levee Board" was fast becoming a household phrase. Still perplexed by an onslaught of conflicting and illogical data, I continued to read and read. I did not yet know that the flooding was due directly to mistakes that engineers with the Army Corps had made in the 1980s.[99] But I did know that focusing on the maintenance folks seemed wrongheaded.

That same month, Congress designated twelve million dollars for a coastal Louisiana hurricane protection study.[100] But there was a stipulation: some members of Congress wanted Louisiana to establish a single state entity to, going forward, act as local sponsor for hurricane projects in coastal Louisiana, which included New Orleans.[101]

Before the 2005 flood, nearly two dozen different levee districts operated as local sponsors for the federally built floodwalls and levees in south Louisiana. In the Greater New Orleans area, there were five: Orleans, Lake Borgne, Jefferson, West Jefferson, and Algiers. The Army Corps is the federal sponsor for all of them. To receive the twelve million dollars, Louisiana was forced to agree for local sponsorship power to be placed in the hands of a single state agency that would report to the governor. It is not stated why this stipulation was placed on the twelve million. It is possible that the members of Congress did not understand that contracts for designing and constructing the hurricane levees are 100 percent controlled by the Army Corps. It is possible they did not understand that the local levee districts only do maintenance of completed floodwalls and levees. (The local officials also must collect local taxes to pay for the maintenance as well as pay 35 percent of new construction by the Army Corps.) It looked as though Congress thought the local levee districts were not paying enough attention to flood protection and were therefore partly responsible.

While this twelve million dollars was being withheld until Louisiana created the new state agency, no money was set aside for homeowners to come home and rebuild. No money was set aside for businesses to reopen. Congress held a decidedly suspicious view of the people of New Orleans. The idea that New Orleanians had brought their misery upon themselves was easy to accept and did not require reshaping one's world view. Washington, DC, kept a continuous focus on possible local corruption.

* * *

But one can work and stress only so much. I needed to stop poring over news stories twenty-four seven and get some exercise. I needed to find an environment that felt normal or as similar as possible to life back in New Orleans, so I applied for a job at the City Club at River Ranch in Lafayette. The manager hired me on the spot because I was a certified fitness instructor with fifteen years of teaching experience and because their membership had swelled to bursting with evacuees just like myself. They were doubly pleased that I agreed to teach a senior class of ladies.

"When can you start?" they asked.

I started the next day. How good it was to be doing something that made me feel—even for just sixty minutes—like I was back home! The music was the same, and the room was similar with its mirrored walls and shining wooden floor. For one splendid hour, I could exercise and not stress and worry. And I made a little money, though not much.

I had also brought my tennis racket and gear and had no trouble finding other ladies who wanted to play tennis. Lafayette was a "tennis town." Whenever I called someone and said that I was an evacuee wanting to play, they would ask when and where. They didn't care what my husband did for a living or where my kids attended school. Touchstones to reality, like teaching fitness and playing tennis, were important to keep me sane during those long days of reading and research.

I also woke up each day and asked myself how I could be a better mother or a better wife. Each day, I tried to do something special for everyone around me. I even tried to be a better master for my dog. Focusing on those around me took my mind off the naked realization that my world had been turned upside down. All my plans for the coming autumn were not going to happen—perhaps not for the entire year.

Focusing on family and friends also kept my anger from boiling over since I had determined that the 2005 flood was not a surprise natural disaster. Every day, I managed to speak to at least one girlfriend, which was not easy to do with so many inactive cell-phone towers. It took at least ten tries to reach someone by cell phone. To make contact, you tried, again and again, until you got through.

I finally reached my friend Debbie Friedman, who owned a clothing boutique on Magazine Street. She related how she was distraught that she had to devote her time to taking inventory of any items that could be salvaged instead of helping her friends get through this difficult period. Another close friend told me that she had moved with her two sons to her parents' home in Dallas, but she was also separated from her husband, a doctor who remained in New Orleans to work. Yet another close friend confided that, in mid-October, her mother had passed away in a nursing home in Baton Rouge. She was convinced that her mother had died from the trauma of being relocated away from her home.

Few days went by that I didn't say out loud how fortunate we were. I could be with my husband, my son, and my dog every single day. Our home and my son's school didn't flood. Both sat atop the same stretch of natural Mississippi River levee that had been built over thousands of years. My husband's business was virtually uninterrupted by the 2005 flood due to meticulous planning long before the wind and storm surge arrived.

After the 2005 flood, everyone in New Orleans revised or created their hurricane plan. Too many people bitterly regretted their decisions to shelter in place, like my friend Jack Davey, a mechanical engineer who lived in a region of New Orleans that historically didn't flood.

Jack had told me, "My wife and I had always stayed. We had a generator that was supposed to last a month. But the generator didn't last a week. It was the stupidest decision I ever made in my life."

Jack and his wife survived. But, just like Renee and Harvey Miller, they survived because they were lucky.

* * *

The Millers went to counseling together the first time. Renee didn't feel that she needed any more counseling after that, but Harvey continued to go every three days. FEMA denied payment for the sessions.

* * *

On November 4, the Army Corps issued a press release, announcing that the chief of engineers, Lieutenant General Carl Strock, had commissioned an Interagency Performance Evaluation Taskforce (IPET) to study the performance of the hurricane protection system in New Orleans and the surrounding areas.[102]

The White House did nothing while the Army Corps—the organization responsible for the flood protection's performance—convened and led an investigation of its own work. Inexplicably, neither Louisiana's governor Kathleen Blanco nor the Louisiana congressional delegation protested such a clear conflict of interest.

All was quiet even while Steve Ellis (Taxpayers for Common Sense) and Scott Faber (Environmental Defense) howled in protest. They wanted "to see some sort of independent federally authorized commission look into the levee breaches, in addition to the [Army] Corps."[103] And, with 500,000 families displaced from their support base (family, neighborhood, and place of worship), citizens could not collectively recognize the travesty or do anything to stop it.[104] They were literally consumed with worrying over life's most basic things.

Chapter

4

The Face of the Monster

On the morning of Friday, October 28, I awoke feeling better than I had in eight weeks. I was going to spend the morning and early afternoon doing something I loved: playing tennis. The familiarity of a tennis court was highly welcome in this new world, post the 2005 flood. All tennis courts are basically the same; in fact, they are all built facing the same direction relative to the sun, a perfect touchstone to normalcy.

This was alumni weekend at the University of Louisiana in Lafayette. One of the planned activities for the alumni was a mixed-doubles tennis tournament. A few days earlier, the planners realized that they were short one female player. One of the planners knew me and remembered that I was always on the prowl, looking for a tennis match to play. She invited me, and I promptly accepted.

The day was glorious with temperatures in the sixties. After the initial introductions, the conversation gradually shifted to my being an evacuee, and I launched into my researched but short script on why New Orleans flooded. The levee failures were due to poor design and construction by the Army Corps. I closed by saying that, had the levees been properly built, there would have been little more than some lost shingles and soggy carpets.

What happened next changed my life more than the 2005 flood alone could have.

My male tennis partner, along with the male tennis player on the opposite team, became instantly angry. They told me that there was nothing wrong with the levees and that the hurricane was a huge storm. They went on to say that New Orleans was a "city below sea level," and we should not expect any special help.

At first, I said nothing because I was shocked. I had grown accustomed to being treated as a bit of a nuisance in Lafayette due to the sheer number of evacuees who had temporarily relocated there. Traffic was snarled at all hours of the day, grocery stores routinely ran out of food, gas station lines were long, and so on. Until that moment, I had felt tolerated, but I had never felt unwelcome. And I certainly had never been told that I, and evacuees like me, deserved our misfortune. Shaking with anger, I walked to my tennis bag and pulled out my car keys. I walked back to the two men who had moments before berated me and my fellow New Orleanians.

I held up my car keys and said, "I am a victim. If you don't apologize right now, I will leave."

Had I left—as I was obviously prepared to do—I would have caused a serious disruption to the tournament.

"I'm sorry!" my partner said quickly, but the deed was done. My eyes were now wide open to the misinformation, and therefore the impossibly large amount of resentment toward survivors of the 2005 flood. These two people, both alumni of this institute of higher learning, lived just hours away from New Orleans. If they had their facts so wrong, imagine how confused people must be if they lived in New England or Chicago or California? I suspected that these two people were not alone in their opinions. Thinking back to the way the folks in Lafayette appeared to tolerate us, I could now see that there might be a similar feeling lurking just beneath the surface.

My tennis partner had apologized, which required me to stay and play the tournament. To my partner's credit, he was on much better behavior after that. But all day long, my mind spun with the revelation that 80 percent of New Orleans residents lost everything—or almost everything—and they were being blamed for it. They were also considered stupid for living there and undeserving of help. Something had to be done!

* * *

The next morning (Saturday, October 29), I cleared my head by leading an invigorating group-exercise class at City Club. Even then, physical exercise always had a glorious ability to turn big problems into smaller ones. Problems didn't disappear completely, but they always seemed smaller after exercise.

After the class, I drove home in the usual bumper-to-bumper traffic in the overcrowded city and found Steve soaking up to his chin in the bathtub after a grueling tennis match. I told him I wanted to join others in an effort to show the nation that New Orleans residents had been unjustly thrown under the bus. I couldn't bear to sit idle, knowing that, behind all the talk of wind and water, the Army Corps was hiding. I believed that there had to be other people who had already drawn the same conclusion. How could I possibly be alone? I started making phone calls and sending emails.

After looking for three solid days, I came up empty handed. There was no one doing the work that I was burning to do: to right the misinformation and to bust the myths.

"Well, then, I will just have to lead the effort myself!" I said aloud in the empty kitchen. "All I need is a spokesperson."

I thought perhaps I could find a famous celebrity who was born in New Orleans and would be interested in helping me. I attempted to reach Ellen DeGeneres, who was born in the New Orleans suburb of Metairie and found out how difficult it can be to get in touch with a celebrity, especially a television star.

I began to understand that, if I wanted to lead an organization devoted to bringing the vetted facts to every household in America, I could not wait to find a celebrity spokesperson.

Stanford and I discussed this that night over a supper of stewed hen, rice, and gravy. I explained to Stanford that there was no time to find a spokesperson.

"I will just have to be the spokesperson myself until I get a celebrity," I told him.

We talked about how we would create an organization to explain the real reason that New Orleans flooded so badly. We needed a name, a mission, and goals. Stanford said that, if I would work on that portion of the project, then he would design and create a website. We decided that we should use the word "levee" since that was primarily where the confusion about the flooding lay.

While I scraped the dishes and loaded the dishwasher, Stanford did an online search for any URLs that contained the word "levee." A true testament that levee breaches were unheard of before August 29, 2005, is the fact that so many URLs containing the word "levee" were not yet taken! Levees and levee breaches were not yet part of commonly used language. On that cold, clear, starry night in early November 2005, we could have almost any URL we wanted.

We kept things simple. We chose "Levees.org" for our URL. The name of the organization would be the same as the website—something that turned out to be effective in an increasingly digital world.

* * *

The Berkeley team had a budget of less than $250,000.[105] In contrast, the IPET (Interagency Performance Evaluation Task Force) budget was in excess of twenty-five million dollars.[106] But since funding for both was federally sourced, the Army Corps had earlier agreed to cross-share field data and lab results. So Dr. Seed formally reached out to Dr. Mlakar, the designated point of contact, to request the cross-sharing in order to fill in gaps. Initially, Dr. Mlakar said that the data would be quickly forthcoming. Then, he said that it would take a bit longer. Ultimately, Mlakar told Dr. Seed, in no uncertain terms, that he and his team "were never going to see those data."[107] Gradually, the Berkeley team realized that the IPET, which had been described by General Strock as a wide range of experts with two different sets of fact-checking, peer-review groups,[108] was in fact the Army Corps investigating itself with the help of preselected consultants.

By November 3, the Berkeley team and the members of the ASCE team had completed their initial report. Though the two teams had earlier agreed to combine their efforts, it was written mostly by the Berkeley team. The team presented their testimony to a highly engaged Senate Committee on Homeland Security and Government Affairs hearing titled, "Why Did the Levees Fail?"[109]

This particular hearing was convened in response to Michael Grunwald's *Washington Post* article, which heavily criticized the Army Corps. Dr. van Heerden was also an expert witness, and he provided testimony that was similar to the Berkeley team. Both focused on the 17th Street Canal; it was not overtopped but rather that the canal's steel-sheet pilings in the floodwalls had failed—a potential design flaw. Furthermore, the canal's floodwalls had breached between four and five feet below design specification. Described in layman's terms, the floodwall failed when water pressure was about half what they were designed to hold.

At this final statement, Dr. Mlakar blanched and urged caution in jumping to conclusions. (Mlakar had told CNN that the canal wall had failed due to the "awesome force of the storm" and that water had risen to the top of the floodwall and coursed down the other side like a powerful waterfall.[110]) Sen. Susan Collins (R-ME) shrewdly recognized that the Army Corps spokesperson was being slow about providing anything useful.[111] Senator George Voinovich (R-OH) was present for this hearing, and he would become a key figure in testimony five weeks later about the same drainage canal wall.

* * *

At this same time, some participants at the original "secret meeting" in Dallas were busy filing a bill with the Louisiana legislature. The bill, ostensibly crafted by State Senator Boasso, was first introduced in a package of bills on October 20 and focused on the Orleans Levee Board. On November 11, when Boasso's Senate Bill 95 was formally introduced in the legislature—and when I was able to read it—I was, as usual, perplexed.[112] It conflicted with information that I had found in the September GAO report; that the Army Corps was tasked with designing and building levee protection and the local Orleans Levee District was responsible for maintenance of completed structures and for operation (e.g. closing gates when hurricanes approached).[113] It seemed to me that, if the levees broke, one should look to the architect and to the contractor, which, in this case, was the Army Corps. To me, blaming the local levee officials was like blaming the janitor if a building fell to the ground. Furthermore, I had seen no stories documenting that the Orleans Levee District personnel had performed their levee maintenance improperly.

At this exact time, Jim Letten, US Attorney for the Eastern District of Louisiana, told a reporter with the Associated Press that he would pursue tips that he had received about corruption relating to building and maintaining the levees.[114] The reporter added, "Local agencies handle most of the building and maintenance of levees."[115] This statement was wrong. The local agencies controlled none of the building.

This news article was circulated coast to coast and nurtured what the American people already believed. Fingers continued to point toward local officials and away from the federal government—away from the Army Corps. The same Associated Press article closed by pointing out that State Attorney General Charles Foti and Orleans Parish District Attorney (DA) Eddie Jordan were also conducting similar investigations of their own.

When the media reports that federal, state, and local DAs are announcing an investigation, that is what is remembered. But, as famously stated by reporter Megan Carter (played by Sally Field in the 1981 classic *Absence of Malice*), the government does not tell the media that an investigation has been discontinued.

Sure enough, the investigations by Letten, Foti, and Jordan were eventually closed with no indictments and no press. But the damage inflicted—at a time when the story was under intense public scrutiny—was incalculable.

* * *

It was now mid-November, and Louisiana politicians were getting an earful from Boasso. He pinned blame for the flooding squarely on the Orleans Levee Board while failing to provide documentation or proof of why.[116] In the words of a popular political science professor at LSU in Shreveport, "Boasso gave an emotional speech, addressing several senators by name, by recounting personal anecdotes about the flooding in St. Bernard Parish, argued how consolidation of levee boards could be the only solution to improving flood protection."[117] Boasso wanted to get rid of the Orleans Levee Board by consolidating it with others (the Jefferson and Lake Borgne districts).

By this time, Jim Huey had resigned as Orleans Levee Board president and was replaced by Mike McCrossen. The booming-voiced McCrossen, insisted that he had plans to improve the way the Orleans Levee Board operated in light of the levee failures.

But the Business Council did not want improvements. They wanted wholesale reformation of the way the Orleans Levee Board members were selected. And they seemed to want the current members, including McCrossen, booted off the board pronto. On November 16, the Business Council issued a press release—a letter written by Jay Lapeyre and cosigned by thirty-six others, addressed to Governor Blanco. There was a smattering of university and organization leaders, but the list was overwhelmingly wealthy business owners.[118] The letter again painted a picture of an incompetent Orleans Levee Board that paid too little attention to its levee-maintenance responsibilities. The letter spoke of "drowned and near drowned citizens and businesses seeking immediate change at the Orleans Levee Board."[119] Lapeyre is quoted as saying "the future of New Orleans and the entire area depends on it."[120] Of course, Lapeyre, et al., are referring to State Senator Boasso's bill, which operates under the assumption that New Orleans officials are to blame for the 2005 flood.

The day after the press release brought another salvo. The Business Council ran a full-page advertisement, costing up to $10,000, in the *Times-Picayune.*[121] John Georges, CEO of Imperial Trading Company (the Port of Louisiana's leading distributor) ran expensive televised ads paid for by him and featuring himself. "Let's put experts not politicians in charge of our levees, clean out the patronage, and build a levee and storm system that protects all of us from flooding and hurricanes."[122]

The Orleans Levee Board officials—eight men and women—were considered instantly guilty in an environment of fear, chaos, and suspicion. And it's easy to understand why. The assumption that the Army Corps was faultless and that the local officials were corrupt was easy to believe and accept. But there was nothing factual to document this theory. I know this because I had been studying it ever since the power was turned back on. It is possible that State Senator Boasso and the business community were confused. Few people on the planet truly understand the machinations of the Army Corps and its workings with local levee officials. But one thing is for sure: the Army Corps knew. And they said nothing as Boasso and the business community made statements that years later would look and sound ridiculous.

* * *

At this time, the Louisiana legislature was poised to approve the creation of a new state agency that would be the local sponsor for all federally built levees in coastal Louisiana, including New Orleans. Governor Blanco was compelled to comply; if she did not, twelve million dollars of levee rebuilding money would be withheld.[123] Senate Bill 71 (duplicate House Bill 141) was approved in the First Extraordinary Session 2005, Act No. 8.[124]

Meanwhile, the Business Council's ads worked. Boasso's bill passed in the Senate with a vote of 37–0. However, on Sunday, November 20, Rep. Kenneth Odinet (D-LA) employed a parliamentary procedure that prevented further debate, effectively killing the Boasso proposal for the legislative session. For the time being, the business community's initiative that would scapegoat local levee officials, and give the Army Corps a free pass, was ostensibly lifeless. But it was far from dead. Soon it would come back to life in the color red. Literally.

* * *

Throughout the month of November, I worked on the grand rollout for our website. This was a difficult chapter for Levees.org because it required defining what we wanted to accomplish. We had a name and a URL, but we also needed a succinct mission and clear goals. After mind-bending imagining, we defined our mission: to create national awareness that the flooding of New Orleans was a direct result of engineering mistakes at the federal level. While I prepared the rollout, Stanford designed the website and logo.[125] Elegantly simple, the logo is a single, curved, green line resembling an earthen levee with the words "Levees.org" written inside.

It would be reasonable to question my belief that I was capable, with only the help of my fifteen-year-old son, of accomplishing this impossible job. I didn't know that I would soon face huge corporations and federal organizations with in-house counsel, deep pockets, and enormous egos.

The most significant handicaps for me and Stanford may be what prepared us best for the challenges we would soon face. I have already mentioned our next-door neighbor Mr. Bouillon on North Roclay Drive and the magic trick that he had performed on Stanford's orthotic inserts.

But there is another side to that coin. For the previous two years, while Stanford's friends and classmates took PE classes during school and practiced team sports after school, Stanford honed his computer skills. Stanford's inability to participate in sports freed up hundreds of hours—hours that Stanford would use to become quite proficient in website design and devising ways to use the internet to further the reach of our message. It would have been impossible for me to launch Levees.org without my son.

In my case, the handicaps that prepared me to lead Levees.org were my speech and hearing problems. Born with "severe sensorineural loss" (commonly called "nerve damage"), I did not have the audio feedback that other toddlers have.[126] Much of my early speech was incorrect. For example, I said "shicken" instead of "chicken." Correction from adults (and from an irritating older brother) taught me to say "chicken;" however, many of the subtler sounds remained uncorrected, and I retained odd speech. At the age of forty, fed up with the confused expressions of people struggling to comprehend me, I undertook the difficult task of relearning how to speak. The therapy left my jaw tired and aching at the end of the day. But I practiced all day, every day. There was no excuse for not improving. As my speech became more understandable, I found confidence in myself, which is uplifting and empowering.

In the fall of 2005, when it became obvious that leading Levees.org meant that I had to be the "voice" of the group, my new confidence in my speech enabled me to submit confidently to radio and television interviews on a rigorous schedule. But I can also point to other things that prepared me to lead a grassroots group.

My entire career, since graduating from Tulane Business School in 1981 (now the A. B. Freeman School of Business) has been in marketing, with a specialty in copywriting. I had a knack for figuring out how to compel readers into action, which was essential in a digital world with nearly limitless competition. In addition, I had been leading group fitness classes since 1990, which taught me how to project my voice, how to be a role model, and how to inspire confidence with just the right balance of energy and calm confidence.

Since 1985, I had volunteered with the local alumnae group for my alma mater, Mount Holyoke College, serving first as vice president and then as president for fifteen years. With this job, I learned the delicate art of managing volunteers. But I wasn't thinking of any of these things when Stanford and I worked on the rollout of our website. I was thinking only of what needed to be done next. And I realized that there was something we didn't have yet: supporters.

If Stanford and I wanted to change the narrative about the 2005 flood, we needed lots and lots of supporters. So I created a petition directed to President George W. Bush, asking him to keep a promise that he made as he stood in Jackson Square on September 15, 2005: to make the levee system "stronger than it has ever been."[127] I asked Stanford to create a page on our website for the petition. On the day that we would launch our website, we were prepared to ask our family and friends to sign the petition, forward it to their family and friends, and so on.

During this time, I relied heavily on Stanford to guide me in the world of the internet. I did everything he recommended. One of his suggestions was to discontinue using AOL and switch to Gmail. I did what I was told, and I am glad that I did. The switch to Gmail guaranteed that I would have access to every email communication I ever received or sent, something that would become critically important, over and over, in the years to come.

Next we decided to host a kickoff event that would give our grassroots group an official start date. But where to hold it? I planned a trip to New Orleans with two goals: recruit supporters and choose a venue for our maiden event. I had heard that a senior partner with Lowe, Stein, Hoffman, Allweiss & Hauver LLP had rented space on St. Charles Avenue and invited members of the community for a planning meeting. I called in advance to ascertain that internet connectivity would be available in order to display our new website on my husband's laptop, which I borrowed for the trip. The meeting was set for seven o'clock on Tuesday, December 6, 2005. I planned to be there.

* * *

Thanksgiving arrived—the time of year where most everyone relaxes a little bit and spends time with family, without the need to buy presents. But for thousands, getting together would not be possible. The tendency for New Orleans families to live very close to each other had a devastating effect after the 2005 flood. Entire extended families were flooded and needed to move to places far from their own neighborhoods. This meant that extended families were often torn apart by hundreds of miles. Getting together for the Thanksgiving holiday—something usually taken for granted—had become either difficult or impossible.

Here, yet again, my family was fortunate. Our extended family was able to get together that year just as we always had: my family of five, the Jacobs family (my sister-in-law Leslie, her husband Scott, and their two daughters Lauren and Michelle), the Buchholz family (my mother-in-law Sandra Rosenthal and her husband Rogene Buchholz), and, of course, Grandma Rose. Dinner was at my house on Soniat Street, just as it always was.

On Tuesday, November 22, while my family was trekking back to our home in New Orleans to get ready to cook a feast and entertain, an article appeared in the *Times-Picayune* with a title that screamed "Failure of levee merger sparks outrage." The article, written by Frank Donze, listed quote after quote from members of Mayor Nagin's BNOBC (Bring New Orleans Back Commission), the same commission that Karl Rove suggested at the "secret meeting" in Dallas.

In the article, Donze quoted four committee chairs at a special meeting of the mayor's commission at the Sheraton Hotel, the central nervous system that acted as a government annex while so many buildings were unusable. In each quote, the committee chairs—representing public transit, land use, higher education, and big business—all said the same thing using slightly different wording:

1. The current Orleans Levee Board setup breeds corruption.

2. Orleans Levee Board appointments are a plum position.
3. Elimination of political patronage was needed.
4. Orleans Levee Board members should have engineering expertise.[128]

It is instructive to note that the cochairs of the BNOBC expert levee committee were not quoted in the Donze article. The levee committee was comprised of experts in civil engineering and hydrology, chaired by civil engineers Billy Marchal and Bruce Thompson. The levee committee proposed infrastructure improvements, but their lengthy report did not recommend changes to the Orleans Levee Board.[129] Apparently, no one asked the experts whether levee-board reform was needed.

Keep in mind that the reform legislation was written and ready just seven weeks after the 2005 flood.[130] Few were prepared to wrap their heads around all these claims and intricacies. Most families were struggling just to move back into their own homes. Few had the energy or the time to ask deep, hard questions and challenge the myriad chairs of the mayor's BNOBC, who were chiming the same mantra.

But what might be the straw that broke the camel's back in the push for "levee-board reform" was the hysteria. It began with this blood-freezing opening line in Donze's article: " 'Obstructionist' state lawmakers are jeopardizing billions of dollars in federal aid for flood protection by preserving an antiquated, patronage-laden system of levee boards."[131] Billions. Here was something brand new: a threat from members of Congress in Washington, DC. They would withhold levee-building and rebuilding dollars if the Orleans Levee Board was not dissolved.

Oliver Thomas, then a member of the New Orleans City Council—and assumed to one day be mayor—later wrote to me in an email, "We were being held hostage [by Washington, DC]. I remember like it was yesterday!"[132]

Years later, in a phone interview with Donze on September 19, 2017, I asked him if he had seen anything in writing to support his opening statement in his article that DC would withhold billions in funding if Boasso's bill did not pass. He responded no, that there was nothing in writing, that it was just assumed.

"Your article begins with a line that was just assumed?" I asked.

I heard Donze sigh before he replied that those were chaotic times. The newspaper's printing press was in Houma, over an hour away. Deadlines were unreasonable, he said. "It was just assumed that we won't get help if we don't help ourselves."

So, here we were: a city of people—steeped in rot, anger, and mold—being told by the city's most respected media outlet that Washington, DC, would not send us federal aid for flood protection unless we pass State Senator Boasso's bill, with nothing to back up the admonition except an assumption.[133]

This environment of fear, confusion, and near hysteria made it easy to create a false story. People were thirsty for information and ready to listen to basically anyone. This unique environment coupled with less-than-rigorous reporting by the media (the sort to which Donze admitted) made it easy to create a fairy tale.

* * *

After the business community's levee-reform bill failed in the first Extraordinary Legislative Session just prior to Thanksgiving, the group put their minds toward convincing Governor Blanco to convene yet another special legislative session. But, this time, they tried a different maneuver.

According to Jacques Morial (the brother of Marc Morial, a past mayor of New Orleans, and the son of Dutch Morial, another past mayor of New Orleans), the business group hired Logan Branding to create a logo and website for a group of ladies that gave the appearance of rising up organically.[134] The name of the group was Citizens for 1 Greater New Orleans (Citizens for 1). The businessmen hired a lobbyist, Randy Haynie, to assist the group in its mission.[135]

According to Mike McCrossen in an interview on October 29, 2013, James Farwell—advisor to the executive committee of the Business Council of New Orleans—also provided lobbying support to the group.[136] The group's brand included a bright-red "1" and, like the group itself, the petition campaign gave the appearance of being a grassroots effort—as though ordinary citizens had come together and united.

The group claimed it had collected 53,000 signatures in twenty-five days.[137] A New Orleans blogger made an accurate observation when he posted: "Signs, Signs, Signs. Everywhere you look are signs for Levee Board One Voice."[138]

* * *

In Lafayette, Stanford and I were poised for our website launch. On December 4, 2005, using our respective Gmail accounts, we sent our first e-blast. It was pristinely rudimentary. We emailed our family and friends, asked them to "click the link" to our new Levees.org website, sign the petition to President Bush, and forward it to their family and friends. The concept of "sign and forward online" was still new in 2005.

And it worked. Overnight, we had two hundred signatures! Now, we were "on the map." We had a website, a mission, and supporters.

* * *

Two days later (December 6), I drove three hours to New Orleans with plans to recruit more supporters and scope out potential venues for our maiden event. This visit revealed a very different New Orleans from my first visit in September.

As I passed through Kenner and then Metairie, there were blue-tarped roofs everywhere, installed and funded by FEMA at exorbitant expense. But, driving through urban New Orleans, it was clear that wind damage was far less excessive here than on the north shore of Lake Pontchartrain or east in Mississippi. Many badly flooded homes in New Orleans would soon be torn down with perfectly good roofs.

As I approached New Orleans, I saw that one thing had not changed: the prominence of the color gray. Everything was still gray. And so many more trees were dead and dying, especially the magnolia trees. Something else was distinctly not right: the smell. The first visit to New Orleans didn't smell anything like this. There seemed to be a permanent stench everywhere, including the portion of the city that saw no flooding. It was a ghastly combination of mold, rot, oil, rancid food, rotting meat—and death.

My first order of business was to visit the site of the Army Corps headquarters in New Orleans at 7400 Leake Avenue. The enormous, three-story building was built on the Mississippi River levee, near the city's Riverbend neighborhood. It was the only structure in the city allowed on a levee. For the rally, I chose the upriver side of the building where the levee was spacious and grassy—a perfect spot with plenty of parking on the streets nearby.

With the rally location decided, I looked for a fast-food restaurant to get a quick bite to eat before the community meeting. To my surprise, none of them had yet reopened. This was another testament to how spoiled I was living in Lafayette with all of life's conveniences. And so, with a grumbling tummy, I went straight to the meeting with my laptop. I was prepared to display our brand-new website and provocative logo.

Attorney Mark Stein led the meeting. His plan was to run a democratic meeting and let the group decide what it wanted to do. Then he would take over with the "how" and the organizing. (Nine years later, on January 2, 2015, Mark would tell me that he realized during the meeting there was no way he could get a consensus of what the group's focus should be. He told me that, in hindsight, he should have come to the meeting with a well-defined plan and asked people, "Are you in or not?")

On that day in early December 2005, I was filled with hope and expectations. Here was an opportunity to reach a lot of people, many whom I already knew, in a face-to-face meeting. Every seat in the large room was taken.

With Stein moderating, many stood to speak about their concerns. Stephen Sabludowsky—journalist and local blogger of Bayou Buzz—stood quietly at the back of the room. But suddenly, in a loud voice, he announced that he had just gotten back from Washington, DC, and he had learned that most of the members of Congress considered this disaster to be our fault.

"They're laughing at us," he said.

"But we didn't build the levees," I said under my breath. I would have my chance to talk, and I waited for my turn.

One community member spoke of the difficulty of reaching the Small Business Administration, which was responsible for doling out loans. Another complained that there still was not even a whisper of talk from Washington about emergency funds for homeowners looking to rebuild. A third spoke about how little information was forthcoming from the Army Corps about fortifying the levees. And a fourth pointed out that all these were reasonable and valid concerns, but that to have an effect at the Capitol, she said, you had to be a large group or at least a coalition of small groups. In terms of what was already on the participants' minds, there was very little uniformity in the room.

When Stein gave me the nod to speak, I blurted out that the levees were a federal responsibility and that fault lay with the Army Corps—not with the storm, not with the geography of New Orleans, and not with local officials. I then announced that the grassroots group I had founded was planning a rally at the Army Corps headquarters. And, as part of the rally, we would all lie down to symbolize the number of souls lost from the levee breaks.

I hesitated for a moment before saying more. I had noticed that everyone was staring at me as though I had just sprung a second head. Instead of continuing, I asked if there were any questions. There were none. So I closed by offering to show our website to anyone who wanted to see it on the laptop that I had brought. After the meeting, only one person approached. It was Nancy Marshall, who had two sons in the same schools and same classes as my two sons. But her reason for approaching me wasn't about seeing the Levees.org website. She just wanted to say hello and ask about Stanford and Mark. To Ms. Marshall's credit, she later would run a successful campaign to consolidate seven property-evaluation assessors' offices into a single office, ridding the city of what was considered a corrupt system of residential-property assessment.[139]

The meeting ended. It was still early, so I drove to Rock 'n' Bowl, a unique, live-music venue that combines a bowling alley and a music club. I had been a regular at Rock 'n' Bowl with a group of girlfriends for the previous five years. I had heard that the club was up and running right after the 2005 flood because the business, on a second floor on South Carrollton Avenue, had sustained no flooding.

For an hour, I circulated around the funky club and recruited supporters for Levees.org. Unlike the community meeting earlier that evening, there was far more interest here. Mary Burns, whose parents had lost their home in Arabi in St. Bernard Parish, became a supporter on the spot.[140]

I made my way back to my car in the dark, quiet parking lot and drove home with no lights except those on my car. By far the eeriest part of that trip to New Orleans was going to bed that night. It was so quiet! No streetcar rumbling, no automobiles, no birds, and no insects. I fell into an uneasy sleep and wished that I had brought Chester with me. It wasn't that I felt unsafe and needed a dog's protection. It was the company I wanted.

* * *

The next morning, I looked forward to getting back to overcrowded, bustling Lafayette. I drove down Soniat Street toward St. Charles Avenue. Just as I turned the corner, a prominent, bright-red "1" caught my eye.[141] There, in my neighbor's yard, was a sign that said, "Levee Board 1 Voice." Along the bottom was a URL: Citizensfor1GreaterNewOrleans.com. I made a mental note to look further into this group. Perhaps we could join forces. How good that would be! It's so hard creating something from nothing.

Back in Lafayette, I looked up the URL. The group appeared to be pushing for Senator Boasso's legislation—the same legislation that the Business Council of New Orleans and the BNOBC championed. This group was not directing attention toward the mistakes of the Army Corps. Their mantra was "no more corruption" and they seemed to be blaming local officials for the flooding debacle. This was not a good fit for Levees.org.

I continued with our plans, choosing not to feel deterred by the chilly reception at Stein's community meeting. Around this time (early

December 2005), Stanford and I got a very welcome bit of assistance. Cheron Brylski, the owner of a public relations firm, contacted me to tell me that she loved our website and offered to send e-alerts to her database about our group and mission. This was our first experience working with a professional who wished to see our group succeed. Over the next ten years, over and over, help from people like Cheron would strengthen and guide Levees.org in achieving its mission.

* * *

On November 17, 2005, Daniel Hitchings, director of Task Force Hope for the Army Corps, testified before a Senate Committee on Environment and Public Works (EPW) hearing in Washington, DC. Under questioning from the members of Congress, which included Senator David Vitter (R-LA), Hitchings said that he could not answer questions because independent teams were in the process of investigating the levee and floodwall failures and were expected to produce final reports by June 2006.[142] These final reports, he said, would provide definitive answers to the public.

Vitter responded bluntly to Hitchings' refusal to speak about the levee failures, saying, "I find your testimony, even your written testimony, frustrating and inadequate."[143] Vitter had specifically asked for these details and was annoyed to have to wait until the beginning of the next hurricane season.

Despite Senator Vitter's frustration, the Army Corps was freed from intense questioning by members of Congress. This same courtesy was not, however, extended to members of the Orleans Levee Board.

* * *

Four weeks later (December 15), another key congressional hearing took place in Washington, DC. This one was called "Who's in Charge of New Orleans Levees?"[144] In the hearing, the line of questioning from the committee, which can be seen on a video recording by C-SPAN, clearly indicates suspicion of the Orleans Levee Board personnel.[145]

The questions focused on the 17th Street and London Avenue Canal floodwall failures. By this time, it was understood that the failure of these two drainage canals caused the majority of the damage to the portion of the city with the most people, property, and infrastructure. The members of Congress appeared to believe that the Orleans Levee Board officials were responsible.

The chairwoman—Senator Collins—who, in previous committee hearings had been leveling cold, formidable stares at Army Corps spokespersons, now seemed to be taking swings at local New Orleans officials. In her remarks, she related a story from a former president of the Orleans Levee Board, whom she did not name, who had given this description of the annual levee inspections: "They…normally meet and get some beignets and coffee in the morning and get to the buses, and the colonel and the brass is all dressed up. You have commissioners. They have some news cameras following you around."[146] Chairwoman Collins's tone and word choices betrayed a distinct prejudice. Something had happened to swing the pendulum the other way.

The origin of this prejudice was a *Times-Picayune* story titled, "Levee inspections only scratch the surface," which appeared a few weeks earlier. In a report that would later receive a Pulitzer prize, Gordon Russell wrote in the opening line that local New Orleans officials "usually have skipped the floodwalls along outfall canals exposed by Katrina as the system's Achilles' heel."[147] The story described the annual levee inspections as "fairly hasty affairs, with dozens of officials piling onto a convoy of vehicles to drive along the levees, stopping at various points for visits of fifteen to thirty minutes…typically ending early enough for the group to make it to a restaurant for lunch."[148]

An article one week later by environmental reporter Bob Marshall also with the *Times-Picayune*[149] further reinforced this story. Marshall wrote that critics of the Orleans Levee Board describe levee inspections as "cursory drive-bys more about fellowship and lunch than looking for problems… Records show that the inspections were scheduled to end early enough for a taxpayer-paid lunch, costing as much as nine hundred dollars for fifty-seven people."[150]

* * *

Chairwoman Collins finished her opening remarks with this admonition: "Not only must we strengthen the levees themselves, but also we must strengthen the oversight of the entire levee system if we are truly to protect New Orleans from another catastrophic failure."[151] Of course, there were no facts revealed in the hearing that would support the allegation that poor maintenance had anything to do with the levee failures, but the damage was incalculable in light of the climate at that time. Russell's and Marshall's articles had crystallized a common assumption; that local officials were lazy and corrupt—or both.

At the hearing, Orleans Levee Board personnel did their best to explain the details provided in these two stories. Max Hearn, executive director for the Orleans Levee District from 1996–2005 and director of operations and maintenance from 1986–1996, testified in plain English: "The Corps conducts annual inspections of the flood-control structures within the Orleans Levee District's jurisdiction and grades the levee board on compliance. During my tenure as the executive director of the Orleans Levee District, the Corps has always evaluated the district's compliance level as 'outstanding.'"[152]

Gerard Colletti, operations manager for the New Orleans District Army Corps, even supported Hearn by stating, "We've dealt with the Orleans Levee District for many years, and their operation and maintenance has been outstanding. So, from the aspect of cutting the grass, making sure that the levees are in the condition for what they were designed from a visual standpoint [sic], and all the inspections that are done throughout the year, we have felt that they've done an outstanding job."[153]

Senator Voinovich, who had watched Ms. Mittal throw New Orleans officials under the bus on September 28, 2005, asked Colletti, "So, you did not see any problems there?"[154]

"No, sir," responded Colletti.[155]

On the C-SPAN video, Senator Voinovich can be seen laughing and shaking his head. Despite an Army Corps staffer insisting that the Orleans Levee District personnel had done an outstanding job with their levee maintenance, Senator Voinovich just laughed in disbelief. His mind was already made up. Even statements by the Army Corps exonerating the local levee officials were received with chilly disregard, even laughter, and ignored.

The December 15, 2005, hearing didn't stop with the consternation over the hasty drive-by levee inspections. The questions then turned to the forty-one million dollar budget that the Orleans Levee District used to conduct its annual maintenance activities. Chairwoman Collins asked Hearn why no money was used on more sophisticated equipment capable of doing more than just a visual inspection. Hearn replied that he had complete faith in the Army Corps' levee system, which was a reasonable thing to say in 2005 in light of the Army Corps' two centuries of experience and sterling reputation.

Collins pressed further and asked former president Jim Huey why the Orleans Levee District had commercial activities, including an airport and a marina. Huey explained that these assets generated revenue, which was then used to maintain the levee structures. The revenues generated from the assets had the end result of lowering what New Orleans residents needed to pay out of their pockets.

In a face-to-face interview on March 10, 2017, Huey would tell me that, at that time in DC, the Orleans Levee Board members and staffers chose to "just answer questions" rather than take an advocacy stance. In hindsight, they regretted not trying harder to explain their role, which was unique in the United States of America. These appointed officials and staffers lacked the advice and guidance of public relations personnel—unlike the Army Corps. The staffers and officials struggled to explain their responsibilities to the members of Congress, who knew nothing about levees.

In this toxic environment, Chairman Collins kept up the pressure. She noted, "When the Orleans Levee District has more than twenty million dollars on hand, you would expect that some of that money is being spent to improve the inspection process and to improve the maintenance. So that is hard to understand. It is hard to understand that kind of balance not being used for some of these safety-related repairs or to improve your ability to detect problems."[156]

These statements appear to have been prepared in advance of the hearing because they bore no semblance to reality and did not acknowledge the things being stated by the experts sitting a few feet away on this day.

Collins ended her closing remarks by repeating her opening remark that the people of New Orleans must be assured that there is effective and efficient oversight of the levees.

* * *

While the members of Congress were grilling the Orleans Levee District officials, Stephen Braun with the *Los Angeles Times* was doing some investigative work of his own with quite a different perspective. He had interviewed a half dozen historians and disaster experts who noted that, since the 2005 flood, the Army Corps was slow to make public all its levee-engineering paperwork and had still not produced a full record of the internal correspondence that occurred during the previous fifteen years.[157] But Lieutenant General Strock, like Hitchings, remained shy, telling Braun, "A failure is really where a design does not perform as intended. If forces we designed for were exceeded, there may not be a design failure."[158]

* * *

Two days later (Saturday, December 17), was a joyful day for the Rosenthal family. We were going home. Stanford's last day of school was December 16, and we all had agreed to pack and be ready to go home early the very next morning. Our house on North Roclay had sold on the same day that we put it on the market, which demonstrates how swollen with humanity the City of Lafayette was even three months after the 2005 flood.

Several neighbors came to say goodbye, including Mr. Bouillon. Our neighbor across the street brought us a bag of delicious satsumas. We were all smiles as I backed the car out of the driveway with our perpetually overexcited Chester hopping back and forth between me and Stanford. Steve traveled back in his own car with a small U-Haul. On I-10, we journeyed past the alternating sugarcane fields and wetlands until the roadway climbed high into the air, rounding a bend at the start of the Bonnet Carré Spillway. And there, off in the distance, was New Orleans! But this time, we were not driving to New Orleans to visit. We were going home to stay!

Our neighborhood on Soniat Street felt different. Before the 2005 flood, most of our neighbors were friendly, but now all of them were. There was a joyful reunion on every trip to the corner grocery store. Conversation would begin with the obligatory, "How many feet of water did you get?" and the necessary somber head nods would follow. But then talk would switch to the positive: "So, where are the kids going to school? What contractor are you using?" And, of course, "Where are you spending the Christmas holidays?" It was so good to be home!

My son Mark flew in from Denver when college ended for the semester, and my daughter Aliisa joined us a few days later. Never has having the whole family together under one roof, doing traditional things had so much meaning. That Christmas had some distinct differences from years past. It was still very quiet at night, especially since the noisy, jangling St. Charles Avenue streetcar was not operating. We invited our next-door neighbors to our house for cocktails on Christmas Eve because their kitchen was still unusable.

But we were the lucky ones. About half a million people were still waiting to find out if they would have enough money to rebuild their shattered, damaged homes. The editorial board of the *New York Times* must have been paying attention because they put this pain into words for all America to see: "If the rest of the nation has decided it is too expensive to give the people of New Orleans a chance at renewal, we have to tell them so… Our nation would then look like a feeble giant indeed. But whether we admit it or not, this is our choice to make. We decide whether New Orleans lives or dies."[159]

* * *

On December 21, 2005, a provision for twenty-nine billion dollars for the 2005 flood victims was tucked into a defense-appropriations bill.[160] But, the Senate refused to allow drilling in an Alaskan wildlife refuge in order to get the funding, which resulted in postponement of congressional action.[161] Finally, on December 30, Congress passed the first legislation authorizing $11.5 billion in rebuilding money for the coastal states affected by the 2005 hurricanes.[162] The Louisiana Recovery Authority, created in October 2005 by Governor Blanco and headed up by her chief of staff Andy Kopplin,[163] would use this money for housing (including rental housing), infrastructure (separate from what FEMA public assistance covered), and economic development.

* * *

Over the Christmas holiday, volunteers arrived—many of them students—to assist with gutting houses and helping people stand back up again. Mitch Landrieu, who was mayor of New Orleans from 2010 to 2018, made this observation about volunteerism in his 2018 book, "We could not do it alone. Faith-based, national, and college groups streamed in from across the country with supplies."[164]

* * *

After settling into our house on Soniat Street, Stanford and I planned our first event, one that would take us out of the "internet cloud" and give us a solid, tangible presence in the City of New Orleans. I emailed our little group of supporters and scheduled our first planning meeting to discuss how we should structure our rally.

At precisely 1:00 p.m. on January 11, 2006, on the corner of Benjamin and Burdette Streets—across from our preselected rally location—I arrived with flyers to hand out to the scores of interested volunteers I expected to meet. I got out of the car and waited.

Ten minutes went by. Then fifteen. Just as I was getting ready to leave, I spotted a petite, long-haired woman about my age, walking toward me. I will never forget Lindy Silverman, whom I still call my "number one volunteer." She was the first—and only—volunteer to show up that day. We greeted one another, discussed possible ideas for the kickoff rally, and parted ways after promising to be in touch.

I went home and told Stanford that only Lindy had showed up. Stanford told me not to worry and that I had already accomplished more than Barack Obama. Stanford had been reading our forty-fourth president's 1995 book *Dreams from My Father: A Story of Race and Inheritance.* Apparently, early in Obama's community organizing days, he had planned a rally, and no one came at all!

That night, our committee of three—Stanford, Lindy, and I—chose Saturday, January 21, 2006, at 11:00 a.m. for our kickoff rally—the Rally on the Levee—because there were two engineering conferences that weekend. Stanford posted the event on our website, and I started "talking it up" in emails to friends and family. I ordered two hundred yard signs and, with our small but growing database, recruited volunteers to place the signs at strategic traffic intersections all over town.

With the help of Cheron Brylski, we circulated an online flyer with Stanford's beautiful logo, announcing a press conference in advance of the "Rally on the Levee." The flyer stated, "The facts speak for themselves. Our city is destroyed, and hundreds of people have died because the levees failed. They broke because they weren't built right, and the responsibility for designing and building the levees is 100 percent the US Army Corps of Engineers'. America needs to know that our flood protection is a federal program that went awry."

Stanford and I became immediate targets of harassment. Our SUV full of yard signs was keyed, and whoever scratched the vehicle made certain to scratch every single panel of the car on both sides. This was suspicious behavior because so few people had moved back to the city. That same day, I also found a dead bird at the back door of my house. It was a large pigeon with its head cleanly chopped off. I fetched some newspaper, picked up the dead bird, and put it in the trash can. I never spoke about it because that would distract me.

When I closed the trash-bin lid over the headless bird, I also closed a door in my brain.

Monday, January 2, was Stanford's first day back at school. How happy I was for him! It was also my first day returning to teaching fitness classes at Tulane University's Reily Center. We were both so glad to be back with friends and classmates. Inch by inch, we moved closer to our normal life.

* * *

The next day (January 3), I received an email. It was from the spokesperson for the Army Corps' New Orleans District, but I would not know the sender's role until years later. On that day, the spokesperson emailed me through her personal account:

> From: Lauren Solis<leh7666@yahoo.com>
> Date: Jan 3, 2006 10:52 a.m.
> Subject: Rally Rescheduled?
> To: Sandy@levees.org
>
> Good Morning—
> Has your rally been rescheduled?
> I have to admit, I did not see the point behind it anyway, but nonetheless, the First Amendment grants you that right.
> I think you should know that many folks in NOLA disagree with your views and have begun to organize in an effort to stop the dissemination of inaccurate information that organizations like yours put out. In our opinion, inciting anger in the residents of this great city does nothing but divide its

people and hurt real progress for rebuilding
the system.

Regards,
Lauren Solis[165]

Five years later (April 2011), I figured out that Ms. Solis was an Army Corps spokesperson. I did this by matching her personal Yahoo account displayed on her LinkedIn profile with the January 2006 email. Her LinkedIn profile claimed "extensive crisis response experience to include recovery for the Army Corps in 2005/2006."[166] Ms. Solis's profile also revealed that she was educated at the Defense Information School at the University of Northern Colorado. Only members of the military can attend this school. In the months after the 2005 flood, Lauren Solis was national spokesperson for the Army Corps' New Orleans District, providing quotes to the *Los Angeles Times* on October 12[167] and the *Washington Post* on October 22.[168]

In 2011, the exact moment that I figured out that Ms. Solis was masquerading as a 'concerned citizen' was caught on videotape. Reporter John Snell with WVUE FOX 8 Channel 26 happened to be in my office interviewing me for a separate unrelated news show. This discovery could have become a story all by itself. But Snell's boss did not okay the story, even though their cameraman videotaped the exact moment where I connected the dots and figured out who Ms. Solis was: a person in a position of public trust, disguising her identity, pretending that she was not intimately involved with the floodwall failures, and using her personal email to attempt to convince me to alter my mission. I did not know whether this behavior was illegal, but I did understand that it was reprehensible.

On January 3, 2006, instead of backing off and calling off our rally, I asked questions. Steve and I collaborated and, using his email address, we wrote back to Ms. Solis that same day and told her that we would "really like to understand where we are inaccurate, or what we haven't taken into account."[169] Ms. Solis never responded to the email. I would just have to continue to wait for someone to jump from behind a tree with "the truth" that would refute my theory.

Ms. Solis's LinkedIn profile was scrubbed a few weeks after I located it. LinkedIn members can see who has been looking at their profiles, and the federal government can scrub information from the internet that it finds worrisome. Luckily, I snapped a screenshot of her profile before it was removed.[170] The email address on Ms. Solis's LinkedIn profile (the same as the sender of the email I'd received January 3, 2006) was the nail in the coffin. The activity was now exposed.

The harassment, which would later become much worse, was off to a definite start. But rather than focus on the sinister meaning of the dead bird, the worrisome email, and the vandalism, I viewed those deeds as evidence that Stanford and I were right. And, if we were not perfectly right, we must be heading in the right direction. After all, was this not evidence that someone didn't like what we were saying? And, no matter what, it was clear that the Army Corps was watching me and everything I did, right from the very beginning.

* * *

As Stanford and I dealt with the unwelcome attention, Harvey Miller was heading in a more positive direction. In Little Rock, Arkansas, the trauma counselor guided Harvey "out of the funk," as his daughter described it. FEMA did not pay for the counseling, but the Millers were able to pay for it with their medical insurance. Harvey continued to see the counselor once a week.

Chapter

5

The Force to Be Reckoned With

For the first three weeks of January 2006, Stanford and I focused on attracting as many people as possible to our kickoff rally. Stanford created a "How to Volunteer" button on our interactive website, and dozens of people responded. While that was remarkable, this was also a remarkable time in history. It was now more than four months after an American city was 80 percent flooded, and Congress had only now just authorized $11.5 billion for homeowner relief.[171] There was still the process of delivering the money into the survivors' hands. In addition, Mayor Nagin and City Hall were not providing the sort of leadership the city's residents needed. As observed by Mitch Landrieu in 2017, "Mayor Nagin and his team were overwhelmed. Major foundations and nonprofit organizations wanted to put fuel into the recovery, but City Hall under Nagin was ill-equipped to send proposals with targeted needs, captured in clear prose, with well-developed budgets."[172] In other words, a void of leadership in the city was waiting for someone to step into it. Levees.org, under my leadership, utilized this opportunity.

* * *

Mary Burns, the Levees.org supporter whom I had met at Rock 'n' Bowl a few weeks earlier, wrote to me:

> hi sandy, the clock in the home of my mom and dad who lived in arabi stopped at 10:50 a.m. when the height of the water reached it…i am gonna hold that clock during the protest…thank you for giving me an outlet to express my anger and outrage…see ya saturday [sic].[173]

Ed Chervenak, a political commentator with Tulane University, wrote to me and said that he would like to volunteer. I told him what I told everyone—that the most important thing he could do was to come to the kickoff rally and bring a friend.

Stanford designed a handout flyer, which listed facts about the 2005 flood. We handed them out everywhere we went.

* * *

On Friday, January 20, I got my first terrifying view from inside a television studio. Roop Raj, anchor for WDSU-TV Channel 6 (the local affiliate of NBC), invited me to appear live on his noon show to discuss the kickoff rally. I was so nervous that my husband drove me to the recently built, gleaming office downtown on Howard Avenue. The handsome Raj was warm and friendly, but I was still nervous. A staffer pinned a remote microphone to the inside of my rose-colored, corduroy jacket and told me where to look when I answered Raj's questions.

"Do not look at the camera," he said.

As I waited for the cue, I thought about everyone who had lost so much: friends who had lost homes and friends who had lost businesses. Just like that, my nervousness vanished and the camera seemed harmless.

The studio manager called for quiet. And then 3…2…1… Action! I calmly answered all of Raj's questions. This bit of television coverage was another opportunity to further the reach of our message. Two hours later, back home, I called Garland Robinette—WWL (AM) Radio's host—for a prearranged, ten-minute live interview. That night, I worried about one thing: attendance.

"Will it be a flop?" I thought. "Will I be all alone out there on the levee, just like that day two weeks ago when only Lindy showed up?"

Another nagging worry was the weather. Forecasters were predicting a 90 percent chance of thunderstorms one hour before our rally was scheduled to start.

* * *

January 21, 2006, dawned misty and gray—perfect for our rally. Fog flowed eerily off of the Mississippi River over the top of the sturdy levee and onto the grassy hill where we set up our rally. The temperature was cool but not cold. Our rally location was sliced in half by a bike path, which ran diagonally up the levee. As participants arrived, we invited some of them to walk up the bike path and back down again. I stationed my friend Debbie Friedman at the top of the levee to be a "human pole" and a point for the rally attendants to turn around and walk back down. This gave the event some steady motion.

My worries about attendance proved to be wasted energy. More and more cars pulled up, and more and more people kept coming. Stanford, remarkably good at photography, snapped away. We could have gotten a generator, a sound system, and a podium, but I preferred to keep the feel of the event at a more grassroots level. The only prop I used was a battery-operated bullhorn. I dressed simply in a

T-shirt with our logo on it, rugged shoes, and plain black pants to communicate that this group was for everyone.

For fifteen minutes, we voiced prearranged chants to the delight of the videographers: "N-O-L-A, make our levees safe today!" Then I pulled out a short speech. After starting with a moment of silence, I read my statement, which said:

- The levees failed because they were not designed properly.
- Responsibility for the design and construction of the levees belonged 100 percent to the Army Corps.
- Metro New Orleans was not wiped out by a hurricane. It was wiped out by a federal project done wrong.
- It is time for Congress to acknowledge the federal government's central role in the 2005 flood.
- Our mission is education. We are a nonpartisan, nonsectarian, grassroots group.
- Only the passion of citizens powers us. Our mission is to send America the facts about the Metro New Orleans flood, and we will not stop until this information becomes mainstream.[174]

I did not look toward the imposing Army Corps building to my right. I had to stay focused on what I was doing. But Stanford could not resist snapping a photo of the military personnel, who were standing guard in their uniforms. After reading my statement, I used the bullhorn to lead the over-four-hundred-member crowd in an exchange.

"Who designed and built our levees?" I yelled.

"Corps of Engineers!" the crowd answered.

I ended with, "Who flooded our city?"

"Corps of Engineers!" the crowd responded.

At 11:20 a.m., I asked the crowd to turn and face north toward Washington, DC.

As we turned our backs to the gray Army Corps complex, we yelled, "You flooded us! Please help us!"

"Louder!" I yelled.

"You flooded us! Please help us!" the crowd hollered back.

I made one big mistake in planning. I didn't anticipate nearly losing my voice after chanting and hollering through a bullhorn for more than an hour. I also did not know that I would be approached by every television station, every radio station, and every amateur holding a video camera for a "personalized" interview. By the time these interviews took place, my voice was scratchy and barely more than a croak. And I had brought no water. But that appeared to be my most serious mistake.

* * *

There was no national press, but, that night and the next day, my name and the name Levees.org were featured in every local media with the exception of the *Times-Picayune.*

The best news story was written by freelancer Allen Johnson, Jr. and featured by *The Advocate.* This excellent front-page Sunday story described why I started Levees.org as told from Stanford's viewpoint.[175]

In all the news stories, the Army Corps was invited to give their comment. But a spokesperson could not be reached.[176]

* * *

In January 2006—perhaps just beginning to understand the enormous amount of data that needed to be reviewed—Dr. Mlakar, representing the IPET participants, requested that the Berkeley team cochairs share their work. In return, the IPET members offered to show their work. Choosing the moral high ground, the Berkeley team cochairs complied. But Dr. Mlakar did not reciprocate the gesture.[177]

What the Berkeley team cochairs would not know until almost two years later is that, on January 31, Larry Roth had submitted a grant request to the tune of two million dollars to the Army Corps' ERDC (Engineer Research and Development Center), the sister agency to the Army Corps.

The wording of Grant W912HZ-06-1-0001 was technical: "…to provide support for the Recipient (Grantee) to conduct scientific and engineering research and review and analysis of the Grantor's research on performance issues concerning the hurricane protection and flood damage reduction system in the New Orleans metropolitan area."[178]

While the verbiage was technical, its meaning is clear to anyone who understands the words "conflict of interest." The ASCE was requesting that the Army Corps pay them two million dollars to perform a review. But no one would know anything about this large sum of money changing hands until 2008 when I received a response from the Army Corps to a request under the Freedom of Information Act (FOIA).

* * *

On January 31, 2006, I received my first communication from the Berkeley team. Dr. Bea, in his signature kindly way, started with a greeting of "Happy Tuesday." He was writing to request one of Levees.org's yard signs. He also sent some documents for me to review. In the short note, he wrote, "Some of us are trying to help so that this will not happen again."[179] Thrilled with the possibility of working in tandem with an independent investigating team, I drove to Prytania Mail Services that very day with one of our yard signs to send off.

* * *

Life for me and my family returned to some semblance of normality. Stanford, with his newly functional legs, tried out for the cross-country track team with a group of close friends. Having been separated from his support base of young people and after experiencing a trauma, he wanted very much to put it all behind him. Like Stanford, I too was happy to be in closer proximity to my network of friends. But the great bulk of my time was spent in growing the membership of Levees.org. While I was labeled early on as being the "anti-Corps" and described as having an agenda to make the Army Corps look bad, that was not the case. My goal was to put the vetted facts about the flooding out there, because the current narrative was unfairly disparaging to the people of New Orleans. After all, they had already been traumatized enough.

We continued to gather more and more petition signatures directed to President Bush, which had a two-pronged effect. The petition was an educational tool, and it was also an excellent grower of our support base. The Rally on the Levee was critical in putting Levees.org in the public consciousness. During the week following the rally, I was offered three radio-show opportunities. On Tuesday, January 24, I was interviewed by James R. Engster with *Talk Louisiana* on WRKF, the National Public Radio (NPR) affiliate in Baton Rouge. On Thursday, January 26, I had a half-hour interview at the New Orleans-based KKAY 1590 (AM). On Friday, January 27, Alex Stone, correspondent with ABC News, interviewed me from Los Angeles.

* * *

I discovered that not all local media, and definitely not national media, was eager to write about Levees.org's message. For example, on Saturday, February 4, I got a call from a reporter with a national CBS News station. He peppered me with questions, ending with this: he wanted to know why Nagin, Landrieu, and Vitter weren't screaming about the Army Corps' mistakes. He said that it could mean lots of money coming our way. I replied that, while I don't speak for the mayor and our two US senators, I could certainly speak as a citizen and resident of New Orleans. I added that I could show him data to support everything I said. The reporter hung up and did not report on Levees.org or our rally.

I learned very early not to judge the success of our events or the soundness of our logic by whether or not the national media reported on it.

* * *

Throughout the highs and lows, I made certain to get physical exercise and some sort of social life with my girlfriends. These connections to my "old life" continued to make my days more normal in a dark world where members of Congress moved slowly when releasing funds to rebuild. The tennis courts at beautiful oak-draped Audubon Park in uptown New Orleans were ruined after being used as pitching grounds for tents during search and rescue. But there were indoor courts at the Hilton Hotel where I played often with my friend Debbie Cobb. Debbie's husband was attorney James Cobb, who was famously defending an elderly couple, the owners of St. Rita's Nursing Home in St. Bernard Parish where thirty-five residents drowned.

Also, during these long, uncertain days, I kept up with a monthly lunch bunch of Mount Holyoke graduates, which I had been organizing for several years before the 2005 flood. We met on the first Friday of each month. On February 3, our little group met at Lebanon's Cafe on Carrollton Avenue, an area that missed flooding by only a few blocks. For that reason, this area was playfully called "beachfront property." I met up with two young alumnae, and we shared hummus and pita bread along with hot, heaping bowls of red-lentil soup. In hindsight, I am certain that both the physical exercise and the "down time" with girlfriends kept me from reaching burnout in those early, scary days when I often felt that I had no idea what I was doing.

* * *

On February 6, I received an email, from a military email address in Iraq, from Travis L. Costanza. He was appalled that I was blaming the Army Corps for the failed steel sheet-pile floodwalls.

"ARE YOU CRAZY," he asked in capital letters.[180] He went on to make the following allegations:

- He claimed to have seen some of the breaches, as a member of the Louisiana Air Guard out of Jackson Barracks.[181]
- He claimed that all the levees "were engineered as the technology allowed them some forty years ago when they were built."[182]
- He blamed Congress for "not funding the Army Corps with adequate funds that should be distributed by the local governments."[183]

In February 2006, I did not wonder why a trooper in Iraq would be writing to me. Just as with Ms. Solis—the Army Corps spokesperson who disguised her identity—I had no reason to believe that this email wasn't authentic. I responded calmly with my usual mountain of data.

Costanza wrote back and asked, "Did you know that the levee-board system is not composed of engineers, but a bunch of bubbas that know nothing about engineering?"[184]

This final email showed that the trooper had run out of talking points. Constanza did not deter me; instead, he gave me another opportunity to hone my message.

* * *

At 7:02 a.m. on Tuesday, February 7—two weeks after the Rally on the Levee—I received my second communication from a member of the Army Corps of Engineers. Like the email from Ms. Solis a month earlier, it was also sent from a private email address. It was from Dave Schad, a Public Affairs officer for the Army Corps' Task Force Hope—the people who were assigned the job of doing the emergency rebuilding of the hurricane protection.

Schad wrote, "Sandy—does your organization have a contact phone number?"[185]

I responded. Later that day, Schad left a phone message, identifying himself and asking me to call him. This was my first official contact with the Army Corps. I called him back pronto. Schad was amiable and chatty. He explained that, for several days, Dan Hitchings, director of Task Force Hope, had been looking at our Rally on the Levee yard signs. Hitchings had decided to invite me to his office, along with three other people of my choosing. We would be welcome to ask anything we wished.

I was surprised. I had viewed the organization as military-styled, with no intention of engaging me—ever. Schad followed up—this time with an Army Corps email address—at 5:20 p.m. with potential meeting times. A date was set for 11:00 a.m. on Wednesday, February 15. To accompany me, I selected my husband Steve and Jack Davey. I also invited Steve's first cousin Lisa Brener, an attorney.

* * *

On February 15, after showing our photo IDs, Jack, Steve, Lisa, and I were allowed through the barbed-wire fence, onto the federal property, and into room 341—a dark, wood-paneled meeting room. In a few moments, Hitchings came through the door. He was a middle-aged, unassuming gentleman, who tended to look down when he spoke. But his calm, friendly voice gave the appearance that he was attentive.

I was prepared to hear a preplanned speech and watch a PowerPoint presentation, but that did not happen. Instead, we conversed for seventy-five minutes during which time I wrote copious notes. We asked question after question and Hitchings answered them all.

As soon as I got home, I reviewed all my notes while they were fresh on my mind. I then prepared an email to send to our supporters. But first, I sent the communication to Schad, because Hitchings had requested that he review our mailing and give it his approval. This was even better, as far as I was concerned. Two days later, my write-up came back from Schad with apologies for the delay. Hitchings said that our letter was fine, and we could send it to whomever we wished.

I offered the letter as an exclusive to Stephen Sabludowsky of *Bayou Buzz* as a reward for giving Levees.org some earlier press on his blog. The letter is long, but I will reproduce the most important details here:

> **Sandy Rosenthal:** There is confusion about who is responsible for what regarding our flood protection in Greater New Orleans.
>
> **Dan Hitchings:** It's very simple. We are accountable to Congress and to the American people for the final product.
>
> **Sandy Rosenthal:** What did the Orleans Levee Board do wrong?
>
> **Dan Hitchings:** There are many investigations on going at this time, and

> the findings are not complete. But I am not aware that they did anything wrong that was significant.[186]

If one could look into a crystal ball and see the future, Hitchings's words—that local officials had done nothing wrong of significance—would be found true. But it would be nine years before Hitchings's words would become vetted, confirmed fact. His words would soon be refuted by upper-level Army Corps officials, the ASCE, and Donald E. Powell, President George W. Bush's reconstruction czar. But, for just a little while, we believed that we were victorious.

* * *

At the state capitol in Baton Rouge, Governor Blanco and Andy Kopplin were seething over the unfair way that Congress had capped housing assistance for Louisiana.[187] In response, they set off for Washington, DC, and met with President Bush's chief of staff to debate the cap.

"Why do Americans on one side of the Pearl River get more for their hurricane-damaged house then those on the other side who were affected by the same hurricane?" asked Kopplin.[188]

After this howling, President Bush announced on February 14 that he would request Congress to give Louisiana another $4.2 billion of community-development block grants or CDBGs.[189] It was over this grant issue that Governor Blanco initially agreed to a special legislative session.[190] Subsequently, the governor also agreed to support the levee-board reorganization touted by the business community.[191]

I was dismayed that the Citizens for 1 Greater New Orleans campaign ignored the role of the Army Corps in the levee breaches. On the other hand, at that time I felt that changing the way the local levee board was selected would not cause harm. So, I decided to support the levee-reform campaign and I posted this fact on our website with a link to the Citizens for 1 website. On February 1, I sent an email to our expanding database in which I reiterated that we cannot blindly trust the Army Corps to protect residents from flooding, and I encouraged them to attend the Citizens for 1's rally supporting State Senator Boasso.[192] My email included a link to their website.

I closed with this question: "I have already made my reservation. Have you?"

* * *

On the day of the Red Rally, I spent considerable time debating what to wear; red or Levees.org's signature green. Deciding that I should be visible as the leader of our grassroots group, I chose a green jacket with faux-jewel buttons and paired it with simple black pants. I drove ten minutes to Audubon Zoo's parking lot where one of the chartered buses—paid for by the Business Council—was stationed. While waiting to board, I spotted my friend Nancy Lassen, and we rode the ninety-minute trip with our bagged lunches in our laps. When we arrived in Baton Rouge, there were well over a thousand people—mostly women—all dressed in red. It was an impressive and well-oiled machine. Speeches were made, and, before you knew it, we were back on the bus and heading home, eating our peanut butter sandwiches.

While we rode home, Governor Blanco was in New Orleans giving her opening speech for the Special Session at the convention center. It was the first time in 125 years that the legislature had met outside of Baton Rouge.

Blanco began her speech by referring to the 2005 flood as a "natural disaster" and made no mention of levee-building mistakes by the Army Corps.[193] But she did talk at length about the Orleans Levee Board: "We must improve the oversight and maintenance of hurricane protection levees. Some levee boards have diverted attention away from flood control to corruption and cronyism. The people of Southeast Louisiana want and deserve a single levee board."[194]

The governor was throwing her support behind reforming the Orleans Levee Board despite there not being a shred of credible evidence supporting the movement. Even preliminary investigation results had not yet been released.

But what was probably the biggest instigator to propel the Louisiana State Legislature toward revising its constitution and punishing the Orleans Levee Board was this chilling and false admonition by Governor Blanco: "There is twelve million dollars in federal money we risk losing if we fail to consolidate those levee boards… I'd prefer not to give that federal money to some other state."[195]

It's not entirely clear why the governor told the people of Louisiana that they had to dissolve the Orleans Levee Board or risk losing twelve million dollars of federal money. It is possible that the governor was confused. It is certainly true that twelve million people in federal levee-investigation monies might have been withheld two months earlier if Louisiana did not create a state agency to be the local sponsor for federal water projects in coastal Louisiana.[196] But Blanco had already complied with that directive.[197] That was a separate thing. There was no record of members of Congress—or of the White House—withholding federal funding unless New Orleans's levee boards were consolidated.

There is another possible explanation. Donald Powell was present at the state legislative committee meetings, which were widely covered by the press. He had been observed lecturing the legislature that levee-board reform must happen or the federal government would look dimly at providing any sort of relief.[198] The *Los Angeles Times* fed this particular fire and reported that the Orleans Levee

Board had been spotlighted as an example of the ills of the levee system and board members had been accused of disregarding their safety responsibilities in favor of real estate ventures including an airport, two marinas, and numerous lakeshore rental properties.[199] (Here again, the very existence of assets was framed as "evidence of corruption."[200])

Powell also admonished Louisiana legislators that, although the Bush administration had asked Congress for an additional $4.2 billion to help Louisiana residents whose homes had been destroyed, Congress had not yet approved the request.[201]

* * *

Under this cloud of threats, the legislative session began. On the first day of the First Extraordinary Session of 2006, Boasso introduced SB 8,[202] a near carbon-copy of SB 95,[203] from the previous special session in 2005. While a wee bit of new language was written into SB 8, in substance, the bills were identical.

On February 17, the state's House of Representatives succumbed to the litany of pressures to punish levee-board personnel, namely:

- Ms. Anu Mittal's cherry-picked and misleading GAO testimony before Congress, which lavished praise on the Army Corps and pointed fingers at the Orleans Levee District.
- The Army Corps leaders' refusal to answer questions in congressional testimony until after the investigative studies were completed.
- Army Corps leader Dr. Paul Mlakar's prevention of the Berkeley team from accessing the levee breaches while data disappeared daily.
- ASCE's Larry Roth's prevention of the Berkeley team from speaking to the press.

- The Business Council of New Orleans's long list of accusations toward the Orleans Levee Board in a chaotic environment.
- The *Times-Picayune*'s stories by Russell and Marshall, which wrongly criticized the Orleans Levee District for lax drive-by levee inspections.
- The Army Corps' silence when blame for lax maintenance and drive-by levee inspections was wrongly attributed to the Orleans Levee Board.
- US Attorney Letten's announcement that he was pursuing tips about illegal conduct by the Orleans Levee Board.
- Senators Collins and Voinovich's acceptance of and belief in the plethora of wrong information in the press.

The final straw came in the form of threats, without basis, that Washington, DC, would withhold twelve million dollars in levee-repair money[204] and $4.2 billion in housing relief[205] from Louisiana survivors if the Orleans Levee Board was not dissolved. This claim came from Powell, was echoed by Governor Blanco, and was repeated across the nation—without verification—by the national media, which was anxious to satisfy an extraordinarily engaged public.

On February 17, the legislature unanimously passed a measure to merge southeast Louisiana's levee boards. Prior to that time, a term-limited governor decided which sixteen people in Greater New Orleans would be levee-board members that control how sixty million dollars of taxpayer money gets spent annually on levee maintenance. (Orleans Parish had eight members, Jefferson Parish had five, and Lake Borgne Parish had three.) Now that control was transferred from the term-limited governor to an eleven-member nominating committee with life terms.

"We've just undone more than a century of history," crowed Boasso, who would soon be running for governor.[206]

Three weeks later, the House Appropriations Committee would strip Bush's $4.2 billion of CDBGs, thus making Powell's threat a hollow one after all.[207]

* * *

Feeling the need to decompress, I traveled to Massachusetts to spend a weekend with my sister Melissa and my two brothers, Rusty and Mike. Best of all, I spent some one-on-one time with my sister's two children, Richard, fourteen, and Kathryn, eleven.

When I got back to New Orleans, I felt recharged. I found an email in my inbox from a young man named Vince Pasquantonio. Vince had responded to one of my recruitment emails and asked to meet for coffee. He said that he had experience working in Washington, DC, and wanted to share some insights. The following Thursday, we met at Puccino's on Magazine Street. Vince looked striking with enormous blue eyes and a head of long, magnificent hair pulled back neatly into a ponytail. Not smiling, he shook my hand before we sat down. And he had a great idea.

Vince understood, from his experience with politicians, that we must unite the people of Louisiana behind a common goal: to encourage Congress to reform the Army Corps of Engineers. The people of the city needed to make sure that our Louisiana members of Congress—the Louisiana delegation—stood behind us in these goals, he said. As our Louisiana delegation fought for us in Congress, the people of New Orleans would in turn stand behind them. But there was something that Levees.org needed that it didn't have, he said: a way for our supporters to contact their members of Congress and demand legislation "with a click of a mouse." On March 14, in a cozy café, Vince explained that laws could be passed to make everyone safer from storm surges, and that Levees.org could be instrumental in making that happen.

The meeting went by quickly, and we had to part ways too soon because Vince had to get back to work. But my head was filled with ideas and possibilities. I was intrigued! Such power to influence and educate at the same time. But how would I make it possible to do this with just a click from our website? When Stanford came home from school, I told him about the meeting with Vince and about the mouse-clicking. Stanford said he understood and would get to work on it. Stanford was fifteen years old, and Vince was twenty-five years old, and I was relying on the wits and work of these young men. Time and time again, I would rely on the expertise and experience of others.

* * *

Rep. Jim Moran (D-VA) visited New Orleans in mid-March and left with his head spinning with falsehoods: "There isn't enough space in this article to detail the stories of graft and corruption that were described during my visit. But it is clear to me that, before we make this substantial investment in the region, Congress must ensure that the authority to expend these funds is given to parties who will exercise the authority properly and wisely, with solid monitoring and oversight taking place."[208] This was typical. Local officials were presumed guilty.

On April 26, 2006, attorneys Pierce O'Donnell of Los Angeles and Joe Bruno of Bruno and Bruno in New Orleans filed a lawsuit against the Army Corps for the collapse of the levees in Orleans and St. Bernard Parishes. At that moment, it became clear to me why the Army Corps had been so quiet while major media like the *New York Times* stated that the local Orleans Levee Board should have been holding the Army Corps' "feet to the fire."[209] Now I understood why the Army Corps had been taking it on the chin so well. There was a lawsuit. Anything that pulled attention and blame away from the Army Corps was a good thing for the federal agency.

* * *

Our members of Congress were the people with the power to send money to help us. Soon the levee investigation conclusions would be released, and these same human beings would turn to the Army Corps for information to make their decisions. Not only could the Army Corps control the message about the 2005 flood, they could also hide behind the technical and confusing verbiage of civil engineering.

For this reason, April 11, 2006, was a key day. On this day, I received an email from a civil engineer named H. J. Bosworth, Jr. It seemed that Dr. Bea with the Berkeley team had contacted H.J. and urged him to guide me in the field of civil engineering. This was welcome news because the technical jargon of civil engineering was what confounded me the most. I needed someone like H.J. I responded right away, and we got to work, collaborating on several projects over the next six weeks in anticipation of the Berkeley team report set to be revealed on May 22 at the Sheraton New Orleans Hotel Grand Ballroom in downtown New Orleans.

* * *

In April, the Berkeley team members—one by one—started receiving phone calls from a member of the IPET team, asking if they had a few minutes to "discuss some things." Then, these unsuspecting experts would realize that they were on a conference call with unknown and unseen others and were being pumped for information.[210] Dr. Seed got such a call himself. Subsequently, he warned the thirty-three other Berkeley team members that such calls would be referred to him. The phone calls stopped.

* * *

Later that month, my cell phone rang. It was a Washington, DC, number. When I answered, a man named Garret Graves told me that he was calling on behalf of Senator Vitter. The junior senator had invited me to submit written testimony to the official permanent record of the US Senate hearing before the Homeland Security and Government Affairs Committee taking place in New Orleans the next morning (April 18) in the Vieux Carré.[211] This was splendid evidence that our credibility was growing. Graves followed up the phone call with an email, and I alerted our database about it.

On May 2, 2006, the External Review Panel (ERP) team released their unpolished but completed report. The team went straight for the jugular and wrote a litany of criticisms; the Army Corps had:

- Based the levee-system design on outdated information, resulting in floodwalls being two feet too low.
- Used a model for designing its system (i.e., for a standard projected hurricane), which was too weak, and should have used a stronger storm as a yardstick.
- Designed the system piecemeal over time without broad system-performance thinking.
- Exhibited a tendency to "cut too close to the margin" when building safe structures.

David Daniel, chair of the ERP, wrote that "the result was gross catastrophic failure."[212]

A Public Affairs officer for the Army Corps in New Orleans gave a standard answer when invited to comment: "We are reviewing the report now."[213]

Dr. Bea told the *New York Times* that the ERP's work was "exactly on" and added that the failure to take new information into account, combined with the choice of a flawed standard for setting the design, resulted in a failure in technology's ability to find its way into the hurricane defenses.[214]

At last, the media reports seemed to be heading in the right direction. On May 9, I received an email from Dr. Bea: "hi sandy, i am working on my presentation for the 22nd—and was thinking about including a 'personal' slide that addresses the 'responsibility' issue—hold the corps responsible = YES. i think it is clear now (at least for me) that they carry the major responsibilities for what happened."[215]

Like the interview with Hitchings in February, this email led me to believe that the organization responsible for the levee failures—the designer and builder of the levees, the Army Corps—would be labeled for all the world to see, and that it would be publicly stated by an independent expert that responsibility did not lie with the Orleans Levee Board, the geography of the city, or the fury of the storm. I was so happy! I sent an email to the Levees.org supporters—now almost five thousand strong—and encouraged them to attend the Berkeley team presentation on May 22. In my email that day, I did not mention Dr. Bea's 'personal' slide, but I explained that the Berkeley team would name the entity that carried the major responsibilities for the 2005 flood.

* * *

On May 15, the Berkeley team presented their main technical findings to the National Research Council in preparation for the June 1 scheduled release of the IPET Draft Final Report. At this meeting, Dr. Seed and his wife were dismayed to see that, among the Army Corps, there was backslapping, laughter, and joking. Dr. Seed, a religious man, felt this was inappropriate.[216] Dr. Seed's presentation was a dry, technical study of the floodwall failure strictly at the 17th Street Canal. When he finished, the cheerful banter had stopped. In contrast, one could have heard a pin drop. At that moment, Dr. Seed realized that the Army Corps' New Orleans District was not doing earnest introspection and making an effort to learn from its mistakes. Instead, he realized, there was denial, spin, and obstruction. And there was, to him, a coordinated campaign to partially rewrite history and downplay some key issues.[217] Equally disconcerting, the ASCE—at least at the headquarters-staff levels—appeared in lock step with the Army Corps.

* * *

May 22 arrived. I donned a conservative black skirt, which fell to mid-calf, and a matching jacket with low-heeled black pumps. I arrived early at the Sheraton with lapel buttons for the Levees.org supporters, and signs which we made especially for the event. While arranging the signs and buttons on a white, cloth-draped table, I looked up and saw a neatly dressed man in khaki pants, a light-blue Oxford button-down shirt, and a dark-blue blazer. He introduced himself as H.J. Bosworth. This was our first face-to-face meeting, and I was surprised to see such a young-looking man. I had expected someone much older based on the formal way in which he wrote, and I said so.

"How old do you think I am?" he asked.

"I would guess thirty-eight or forty."

He told me that he was forty-seven. I was thrown off by his full head of thick brown hair, without a trace of gray.

Soon people began to arrive. Many of them were Levees.org supporters, and I encouraged them to sit together. A tall, very thin man with a mop of curly black hair strolled up to me and introduced himself as Ralph Vartabedian with the *Los Angeles Times*. I was glad for this chance meeting because he had written a very damaging article about the local Orleans Levee Board, which had appeared on Christmas Day 2005. The article laid blame for the outfall-canal floodwall failures heavily on the local levee board. I told Vartabedian that I would love to provide comments about the presentation, and he promised to talk to me afterward.

A large group from the Army Corps arrived. They all sat together in the seats closest to the podium. The Berkeley team cochairs arrived early, before the rest of their team. I hovered excitedly around Dr. Bea as he was approached by a reporter with ABC National News. Dr. Bea waited while the cameraman prepared the lighting.

When everything was ready, the reporter got straight to the point. Holding the mike close to Dr. Bea's face, she asked whose fault it was that the levees broke. I held my breath and waited for him to discuss his "personal" slide.

"That will be up to the courts," he said.

I was shocked. Just two weeks earlier, Dr. Bea had told me that, for him, the Army Corps was responsible. I decided to wait until I'd heard the report presentation. Maybe Dr. Bea's "personal" slide was there! There just had to be something in it to assign blame where it belonged.

* * *

The Berkeley team presentation took about forty-five minutes. I had a good bit of difficulty understanding it, but I got the general impression that responsibility for the flooding was "shared" and that everyone had a role in the flooding. I did not see Dr. Bea's "personal" slide, and most alarming—and damning—was this sentence in the executive summary of the Berkeley report:

> The USACE had tried for many years to obtain authorization to install floodgates at the north ends of the three drainage canals that could be closed to prevent storm surges from raising the water levels within the

> canals. That would have been the superior technical solution. Dysfunctional interaction between the local Levee Board (who were responsible for levees and floodwalls, etc.) and the local Water and Sewer Board (who were responsible for pumping water from the city via the drainage canals) prevented the installation of these gates, however, and as a result many miles of the sides of these canals has instead to be lined with levees and floodwalls.

I did my best to smile and show courage. After all, Hitchings, director of Task Force Hope with the Army Corps, had told me that the Orleans Levee Board had done nothing wrong of significance. And Dr. Bea had told me in an email that the Army Corps carried the major responsibility. I held my head high and decided to find out more about this "dysfunctional interaction."

After the speakers were finished, a few reporters asked for my comment, and I complied. Gradually the room emptied. I searched for and found Vartabedian and let him know that I was ready to provide a comment. I waited for him to pull out his pad and pen to take notes. But he just stood there with his arms crossed over his chest. Then, with a crooked smile, he asked me if the goal of my organization was to get as much money as possible from the federal government.

With an inward sigh, I replied "No."

I explained that the goal of Levees.org is to tell the American people the truth about the flooding. I realized that he had no intention of including my comment in his story, so I decided to turn the table and interview him. I asked him if he had backup data for his Christmas Day article titled, "Levees Weakened as New Orleans Board, Federal Engineers Feuded," which ran on the front page of the

Los Angeles Times. He remembered the story and said that his data came from Corps spokespersons. I asked if I could see the notes, and he replied that he was not at liberty to share them.

* * *

Before Dr. Bea left the meeting, Steve and I offered to take him to dinner that night, and he accepted. Over redfish at Ralph's on the Park next to historic City Park, we asked Dr. Bea why he decided not to finger the Army Corps and instead wait for the court system to determine who was responsible for the flooding. He replied that, while he and his colleagues did their investigation in the field, Army Corps spokespersons had told them a consistent story—like it was household knowledge—about how the Orleans Levee Board had used its political connections in Washington, DC, to force the Army Corps to build the inferior system that failed. Dr. Bea believed, at the end of the day, that the Army Corps was responsible; however, he also believed the officials, who served on the Orleans Levee Board in the 1980s, played a contributing role.

I was not familiar with the "political connections" story, but I tried to tell Dr. Bea that the American people will focus solely on the local officials' secondary role—assuming that's true—and will give the powerful Army Corps a free pass.

Dr. Bea disagreed. "People will read all the data and will draw the right conclusion," he assured me.

I was not convinced.

* * *

While we dined, the national news articles flew. CBS News posted an article of which the entire opening paragraph was a list of reasons to pardon the Army Corps: "New Orleans' levee system was routinely underfunded and therefore inadequate to protect against hurricanes according to an independent report released Monday. The report also called for an overhaul of the agencies that oversee flood protection. It took aim at Congress for its piecemeal funding over the past fifty years, and at state and local levee authorities for failing to properly oversee maintenance of the levees."[218]

Of course, the charges of underfunding and improper maintenance were unsubstantiated and false. But the opening lines were catastrophically damaging despite the fact that the rest of the CBS News article was rather good in its reporting. For example, the correspondent Byron Pitts reported that, in several places along the 350-mile levee system, the Army Corps used cheap material (sand instead of clay), and that, according to investigators, the foundation for the levee walls was in some places just thirteen feet deep when they should have been sixty.[219] But these mistakes by the Army Corps were unnoticed because attention was first focused on local officials.

"Infighting between local politicians led to deadly inefficiency," wrote Pitts.[220]

The damage was reinforced the next day in *USA Today*. Dr. Seed was quoted in the national newspaper, saying, "People died because government agencies, from the Army Corps of Engineers to local levee boards, failed to do their jobs properly. Safety was trumped by a desire for efficiency and saving money. Infighting among local agencies and failures in funding by Congress caused years-long delays."[221]

Keep in mind that the Berkeley team cochairs had been barred from the levee breaches. And they were told a story by Army Corps officials that the Orleans Levee Board has used its political connections to force the Army Corps to build the system that failed.[222] One might wonder why the Berkeley team would have reported the unsubstantiated claim if, at the end of the day, the Army Corps were to blame. Apparently, as civil engineers, they felt it was their duty to speak of the human factors involved; their belief, albeit wrong, that the Orleans Levee Board had used politics to force the Army Corps' hand.

Law professor Robert Verchick at Loyola University talked wisely about this issue. "There are other causes to be sure—lax building codes, poor zoning, underinsurance, etc., etc. But it is not unreasonable to emphasize the most direct example of failure. From a moral standpoint, those most directly involved in causing harm bear a particular responsibility. From a practical standpoint, addressing the source of the most direct failure can often lead to the most influential improvements."[223]

But words of wisdom like those from Professor Verchick were lost in the wind as media outlets placed any suggestion of local interference front and center. There was nothing I could do except continue my work with Levees.org. I kept telling myself that there was nothing odd about setting the record straight about an event that killed 1,577 people and forever altered the lives of a million people.

* * *

With help from a Levees.org supporter, Eric Kodjo Ralph, we decided to hold a ceremony on Memorial Day, which fell on May 29, nine months to the day after the 2005 flood. I called Harkins the Florist and ordered a thousand white, red, and pink carnations—differently colored to represent diversity—and invited all Levees.org supporters, asking everyone to wear white, come out to the breach site of the 17th Street Canal floodwall, and "drop a flower into the water" to memorialize those lost. I got a phone call from a local bagpiper, who offered to play at no charge as people were arriving. A supporter named Albert Guidry called me and explained that he was concerned that the Army Corps was doing a lot of construction at the site, and that I might receive some harassment. He told me that he was a big guy, and no one would bother me while he was around. I accepted.

The event was well attended, despite the sweltering heat, and many media outlets were there. Senator Vitter sent a statement for us to read, and Congressman Bobby Jindal sent a representative to read a statement. Then I read my own prepared statement. When I got to the point where I explained that we would drop a flower into the water to memorialize the souls who were lost, I teared up and my voice choked. In unison, the camera operators swung their cameras toward me and took a step closer. I was probably choking for only a few seconds, but it felt like ten minutes.

I finished my short comments, and then ten-year-old Will Hightower dropped the first flower into the water in memory of his grandfather, who perished of a heart attack in his attic on August 29, 2005. Vince sang "God Bless America" as, one by one, the attendants dropped a flower into the water.

I was pleased with the supporter turnout and the news coverage, both local and national. The event was covered by every local television station and media outlet in town except for—as usual—the *Times-Picayune*. Cain Burdeau of the Associated Press did a fabulous piece with photos of our event and a similar event that same morning in the historic Lower Ninth Ward. The story ran nationwide in over a hundred media outlets.

The *Times-Picayune* ran Burdeau's story too, but the news editor deleted Levees.org's event from it.

* * *

Three days later (June 1, at 9:20 a.m.), while sitting at my computer, I read on NOLA.com that the Army Corps was—at that very moment—delivering the results of the IPET Draft Final Report at the Marriott Hotel in downtown New Orleans. The press conference had begun at nine o'clock. I didn't know about the press conference until it had already begun. Apparently, only a select few were invited. I sprang from my chair, threw on a suit, applied my makeup in record time, and drove downtown as fast as could. Fortunately for me, every traffic light was still a four-way stop sign.

I parked at the Sheraton and ran into the hotel and up to the second-floor ballrooms, asking where the Army Corps was holding a press conference. No one seemed to know! Desperately, I asked every person I saw until I found someone who knew. He guided me to a room at the end of a long, dark, narrow hallway. I entered the tiny, quiet room, which was filled to overflowing with men in military uniform and members of the press.

I entered the room at 9:55 a.m., just thirty-five minutes after learning about it at home. When I entered, everyone in the room turned to look and see who had entered. I walked to one of the only two empty seats and sat down. The mood felt grim.

Lieutenant General Strock from the Army Corps was telling the crowd of reporters that there had been a failure of one of the Army Corps' systems. "It was a system in name only," he said.[224] When Strock was finished with his remarks and his dry PowerPoint presentation, which explained the contents of the seven thousand-page IPET report, he opened up the floor for questions.

One by one, reporters from media outlets all over the world approached the mic. When I heard a reporter introduce himself from the *New York Times*, I looked up to see John Schwartz, all five feet, four inches of him. I didn't hear Schwartz's question, but I heard Strock's answer. Strock said he believed that outside influences played a role in the problems of the flood-protection system, though that did not absolve the Army Corps. He was vague, speaking of *mea culpas* and the Army Corps not laying blame on anyone else. Nonetheless, Strock told Schwartz, who reported it in his article later that day, that the Orleans Levee Board had thwarted the Army Corps' plans for gate structures. Strock described the levee system the federal agency had built as a fallback plan. He added that the Army Corps knowingly assumed a higher level of risk due to pressure by local New Orleans officials.[225]

My mind swirled with these accusations—such huge accusations that I knew viscerally had no basis in fact. Having found out about the press conference just minutes before, I didn't have a question prepared, but it didn't take long to think of one. I got in line and waited for my turn.

When I got to the microphone, I looked at General Strock and asked, "As the hurricane was roiling toward shore, did you, or anyone with the Army Corps of Engineers, know that there was going to be so much death and destruction?"

He replied "No," and then told me that I [meaning the people of New Orleans] always knew that this could happen.

Because I was a novice, I didn't understand that I had every right to challenge the commander. I could have said, "No, we didn't. Even the most desperate appeal to residents to evacuate didn't warn that the levees could breach and fail." But I was a novice, and I said nothing further.

I sat in a seat that had just become vacant next to Mark Schleifstein, a reporter for the *Times-Picayune*. I didn't recognize his face, but I recognized the name on his media badge. A year before the 2005 flood, Schleifstein had won a Pulitzer prize for his five-part series titled, "Washing Away," and in June of 2006 was being credited for predicting the 2005 flood.[226] Schleifstein would later qualify that claim by pointing out that no one, including him, knew that the levees would break.[227] I introduced myself to Schleifstein. It may have been my imagination, but he appeared to cringe from me.

I ambled away to listen to the many conversations in the room. I saw that Vartabedian was speaking with Brigadier General Robert Crear, the commander of the Army Corps' Mississippi Valley Division. I nodded my head toward Vartabedian, who pretended not to see me. But Brigadier General Crear turned to me, stuck out his hand, and introduced himself. I took his hand and told him who I was.

"Yes, I know who you are," he said pleasantly. "I have been following your work. You should stay the course because you have more power to change the way things are done at the Army Corps than I do."

As I soaked this in, I felt someone touch my arm. It was Betty Ann Bowser from Jim Lehrer's *PBS NewsHour*. She wanted to hear my commentary on today's events, but she wanted to interview me at the site of a levee breach. We agreed to meet at the 17th Street Canal floodwall breach site at 12:00 p.m. sharp. I arrived on time, and it was hot! Ms. Bowser's heavy makeup glistened in the noon sun, but she bravely withstood the searing heat.

Holding a microphone to my face, she asked, "Are you happy? The Army Corps has just said that they are sorry for the flooding and that they had a major failure. Isn't that what you wanted?"

"Actually, no. We are not pleased," I replied. "The Army Corps has taken responsibility 'at the end of the day' supposedly, but they have shifted responsibility away from themselves, in effect, by blaming local officials for blocking the Army Corps' plans."

"But the Army Corps has said that they had a major failure!" exclaimed Ms. Bowser, clearly exasperated, probably by both me and the heat. "Isn't that enough?"

"No, it's not," I replied. "Until Congress recognizes that the failure of the levees and floodwalls were a federal responsibility, the people of this region will not get the relief they deserve."

I went home and glumly watched the news reports filter in, all of them laying partial, but significant, blame upon the Orleans Levee Board. The story that was repeated the most in other news sources was the article by Schwartz with the *New York Times*, which quoted Strock heavily. His remarks began with talk of the Army Corps "not ducking responsibility"[228] and ended with talk of the Army Corps having ultimate "responsibility for what it builds."[229] But sandwiched between this beginning and ending was a laundry list of reasons to pin blame on others. The lieutenant commander made it clear as glass that he believed outside forces were partly to blame.[230]

Ten days later (June 11), in another article by Schwartz titled, "Too Bad Hippocrates Wasn't an Engineer," there was a quote from Dr. Bea: "Within the corps, complacency and lack of communication paved the way to disaster. But there was plenty of blame to go around. The federal government did not adequately finance the hurricane protection project, and local officials defeated proposals for stronger measures like canal floodgates that would have made their own costs greater."[231]

When Dr. Bea made these false statements, he was fingering living, breathing people (John Ross, Lambert Boissiere, Jerome Dickhaus, Robert Maloney, Steve Medo, Janet Phillpott, Robert Ramelli, and James Smith), all public servants serving on the Orleans Levee Board. Fifteen years earlier, on May 30, 1991, these people had announced to the people of New Orleans, with genuine pride, that the Army Corps would soon build the best protection possible, and the federal government was going to pay for it. These eight members of the Orleans Levee Board had fought for what they thought was best. They could not have foreseen that the Army Corps was going to build floodwalls that were doomed to fail. They could not have foreseen that Dr. Bea and the other Berkeley team cochairs would be bamboozled by Army Corps' spokespersons who, after the floodwalls failed, got straight to work rewriting history. The Berkeley team cochairs drank up the Kool-Aid despite having witnessed the Army Corps hiding critical information and controlling access to data.

* * *

The work of Levees.org was indeed cut out for us. While the world had been told that the Army Corps was responsible "at the end of the day" for the flood protection, it had also been told that local New Orleans officials had a significant role in the disaster.

The challenge was too great, but I didn't know that yet. I was naive enough to believe that, if the American people understood the facts, and if they knew about the Army Corps' terrible mistakes and their subsequent efforts to cover them up, they would see that this disaster could have happened to them too. I continued to lead Levees.org. I continued to articulate the message and drive it, always looking for a novel way to breathe new life into the effort.

Ms. Bowser's *PBS NewsHour* story appeared twelve days later (June 12, 2006). She included some video footage of me, but not my objections to Strock's remarks.[232]

* * *

Despite the pressure from the ASCE's staffers, the ERP report was quite good. For example, it concluded that, had the levees and floodwalls not failed, nearly two-thirds of the deaths would not have occurred.[233] The authors also stated that over 80 percent of the New Orleans area was flooded with "approximately two-thirds of the flooding attributed to water flowing through breaches."[234] The report drew no conclusions about the performance of the Orleans Levee Board.

Berkeley team cochair Dr. Seed felt that it would have been appropriate for the ASCE to present these findings on the same day as the IPET's findings in New Orleans. He felt the report was honest with uncomfortable details about the Army Corps, and it was very readable and accessible to laypersons. But the ASCE headquarters protested that the ERP report was not yet polished enough. Dr. Seed felt that the spit and polish were not important. After all, an engaged world was ready to hear what happened and why. But the ASCE headquarters dug in its heels and postponed the presentation of the final report until June 1, 2007. In Dr. Seed's view, postponing the report was intentional and happened in collaboration with Dr. Mlakar, because the longer it waited, the less public attention it would receive.[235] The final slap was that ASCE headquarters issued a one-year anniversary statement, which stated repeatedly that the Army Corps was not principally to blame for the flooding.[236]

* * *

A few days after the Army Corps' mea culpa on June 1, the $4.2 billion in housing assistance was finally passed into law. The money was expected to be enough to run a homeowner housing program like the one Mississippi was launching, covering rebuilding costs minus insurance and FEMA payments up to $150,000 per homeowner.[237] Ultimately, that still wasn't enough, and in early 2007, three billion more was requested. Senator Landrieu managed to get that money in an appropriations bill in October 2007.[238]

* * *

On June 21, 2006, I submitted an opinion piece to the *Times-Picayune*'s editorial board. In the piece, I stated, among other salient points, that no members of Congress were talking about the Army

Corps's levee failures other than the Louisiana delegation. Annette Sisco with the *Times-Picayune* responded, saying, "Some members of Congress have called this a man-made disaster and one of those was Senator Ted Stevens [R-AK]. We have to pass this time, but thanks for the offer."[239]

I felt the refusal to print was ridiculous. Senator Stevens was just one of 585 members of Congress. As I was new at this game, I did not yet know how to negotiate. I could have sent my op-ed to another paper, such as *The Advocate*. Because I was new at this, I compromised and agreed to change that one point. Sisco said she would run the piece, but the *Times-Picayune* never published it. I posted the op-ed article to the Levees.org website and sent it to our ever-growing database.

* * *

On June 24, I attended the funeral of a drowning victim. Even ten months after the 2005 flood, the bodies of drowning victims were still being found. Mike Patterson worked as a groundskeeper at Atkinson-Stern Tennis Center, a diverse public tennis facility and one of the oldest in the country. Before the 2005 flood, Mike was a daily fixture at Atkinson-Stern with his mile-wide, gold-toothed grin. After the funeral, we gathered with many other tennis players who came to pay respects to Mike's family.

In this city of only about a half million people, it was very possible to know someone who perished or someone who knew someone. No matter who you were in New Orleans, it was impossible to be completely distant from the 2005 flood.

* * *

Communicating the sheer magnitude of the flooding event was a constant challenge for Levees.org. Stanford, on his own, invented a way to do that. Using his computer skills, he took hundreds of photos of flooded homes and created a mosaic. Up close, one could see pixels of flooded homes, but if one stood back from the assemblage of different forms, there was a new whole, an American flag. The plan was to offer it at the first anniversary of the 2005 flood as a fundraiser.

Levees.org spent the summer of 2006 taking Vince Pasquantonio's advice and garnering national support for federal legislation that would reform the Army Corps. We threw our weight—or the little we had—into supporting Senators John McCain and Russ Feingold who, even before the 2005 flood, had been pushing for reforms. Our scattered New Orleans population could be instrumental in making changes to national policy. Our database consisted of one-third New Orleans-area residents and two-thirds from elsewhere, especially Georgia and Texas.

With Vince's guidance, we created a letter on Levees.org's website, calling for Army Corps reform. With a click of a mouse, our supporters could send a letter to their members of Congress, demanding that reform measures—called the McCain–Feingold Corps Reform Amendments—be attached to the Water Resources Development Act (WRDA). This federal legislation could make all people safer because it would change the way the Army Corps handled studies for levee improvements. The law established triggers for independent review of expensive (over forty million dollars) or controversial projects.

So, the campaign began. We asked our supporters to write their members of Congress, demand the reforms, and to contact their friends and family all across the country to do the same. To our surprise, neither of Louisiana's two US senators supported the reform measures. Senator Vitter was most vocally against it, having complained that reforms would slow down the already notoriously slow Army Corps.[240] Senator Landrieu was not so blustery, but she said little beyond that she and her office were examining the issue. In our view, both senators should have been banging a drum in support.

In response, I sent a letter to the editor (LTE) of *New Orleans CityBusiness,* on July 17, 2007, calling out this behavior on the part of our two senators. It was published three weeks later in the August 7th issue. On that day, I received an email forwarded from Senator Landrieu's communications director, stating that Senator Landrieu had voted for each of the amendments on July 19, 2007, two days after I had submitted my letter to the media outlet.[241] That was an excellent outcome.

The McCain–Feingold Corps Reform Amendments to the WRDA of 2007 was passed by a narrow margin of fifty-four to forty-six by the 109th Congress, 2nd Session. Predictably, Senator Vitter voted against it, but Senator Landrieu voted for it. Political observers would credit Levees.org for applying pressure to Senator Landrieu.

* * *

On July 31, 2006, the final Berkeley team report was released. Dr. Seed would later say that he was daunted by the reception. He, an experienced professional, had looked forward to "elegant debates" with the IPET analysis team—people that he perceived as highly qualified in geo-forensic analysis. He expected "nuanced arguments" regarding the details of the failure mechanisms. But the elegant debates and the nuanced arguments that Dr. Seed looked forward to never materialized because the different teams were, in effect, roped off from each other.[242]

* * *

On August 5, the US House broke for the summer without a resolution to mark the 2005 storm that triggered massive floodwall failures in New Orleans and killed 1,577 US citizens. This did not surprise me, because, even though the Army Corps in June had admitted to a major failure of its system, there had still been literally no talk coming from the US House and Senate of how the federal government had failed the people of New Orleans. Meanwhile, the Army Corps-sponsored IPET failed to deliver its promised comprehensive assessment of risk for current conditions by August 1. This was needed so that individuals and businesses could make informed decisions about reinvesting their lives and assets into rebuilding in New Orleans or move permanently elsewhere.[243]

* * *

With the first-year anniversary of the 2005 flood just weeks away, I felt that we needed to observe the event in a meaningful way. Eric Kodjo Ralph, who had given me some good ideas for the "flower on the water" event in May, suggested that we make two "report cards": one for the Army Corps and their post-disaster response, and the second for members of Congress on their post-disaster effectiveness in securing money and assistance for the survivors. Such props—basically two poster boards—cost little and needed just a wee bit of creativity.

We organized a press conference and invited our members of Congress as well as state legislators and members of the New Orleans City Council. Ever vigilant about showing that the 2005 flood was a federal event with significance for everyone in the nation, we chose the Hale Boggs Federal Building-Courthouse for the press conference.

Responses to our invitations came back surprisingly robust. Senator Vitter would appear personally and so would Congressman Jindal. Several members of the New Orleans City Council indicated that they would be there, including Arnie Fielkow and Stacy Head.

* * *

August 22 dawned hot, like all New Orleans summer days, and I dreaded putting on a suit. I would have preferred to dress the way I did at the Rally on the Levee (in a T-shirt and cotton pants), but I felt that I needed to project a more professional image. So, that morning, I donned a taupe-colored suit with a mid-calf skirt and black-lace accents at the lapel and pockets and classic black, low-heeled pumps.

As the podium and sound system was being set up, my stomach was tight with worry. Would anyone show up? Would the headliners—Vitter and Jindal—be no-shows? My worries melted when I saw a widely smiling Jindal with Luke Letlow, his handsome young district director, approaching. Meanwhile, video camera after video camera arrived as well as many members of the press. It was going to be okay! About ten minutes before the event began, a large cloud blacked out the sun offering relief from the heat, but no rain threatened. It was the best you could ask for in August.

So, the press conference began, billed not as a memorial to a natural disaster but as an observance of the worst civil-engineering disaster in the history of the US. First, we unveiled our main prop—a super-sized report card—that the media loved.

Our report to the Army Corps was:

B- for admitting their mistakes

C+ for short-term levee improvements

"Incomplete" for a comprehensive long-term plan

For our US Congress, the grades were:

> **D**- for recognizing the mistakes of the Army Corps
>
> **C**- on Army Corps reform

One by one, we gave each elected official an opportunity to speak. In closing, I invited everyone to view Stanford's mosaic flag which was available for a donation. Everything went smoothly, and I submitted to at least five interviews afterward.

For the remainder of the day, I did radio and phone interviews, including one with the *Washington Post*'s investigative reporter, Michael Grunwald. I had seen and been impressed with his work, and so I felt encouraged when he requested an interview. Grunwald and I talked for thirty minutes, during which time I brought up one of my favorite talking points: If a building fell to the ground, you wouldn't blame the janitor. You would blame the architect, the engineers, and the contractor—all of which are the Army Corps.

One week later, Grunwald's article appeared in *Grist* magazine and it began with my janitor analogy. But Grunwald used it as his own. I was quoted nowhere in the piece. As the leader of Levees.org, I was not afforded the luxury of righteous indignation over Grunwald presenting my ideas as his own. My job was to remain focused on the mission, which is putting the vetted facts about the 2005 flood in front of the American people. So, I sent Grunwald this email:

"I am glad you like my analogy of the levees and a building's janitor. It doesn't concern me that you didn't quote me, but hopefully you will mention—where fitting—that there is an activist group here in New Orleans that gives concerned citizens some education and action points."

He responded, "Thanks, Sandy. I had a quote from you in there—the one about how you're not a weatherman, but you know the hurricane missed you, etc.—but it didn't make it to the final version; we decided to make the entire piece quoteless. And thanks for the janitor idea!"[244]

I responded, "In return for using my idea, please consider—at the fitting time—to mention our activist group."

"You got it," he replied.[245]

* * *

After the dust from our successful anniversary event settled, Stanford and I started to think about getting registered in the State of Louisiana as a nonprofit. Symbolically, on August 29, 2006, we signed the paperwork at the law office of Lisa Brener, my husband's first cousin. Our initial board was tiny, consisting only of me as president and my husband as treasurer. Our concern was more about having something official set up in case our detractors accused us of taking donations and pocketing the money.

* * *

The talk of the corrupt Orleans Levee Board had seemed to die down, but, just as we registered our group name with the state, the rhetoric ramped up again. The Business Council of New Orleans had begun spending hundreds of thousands of dollars for statewide radio, print, and television campaigns to build voter support in advance of the September legislative session.[246]

On August 23, an email landed in my inbox, urging me to vote on Constitutional Amendment No. 3, which would consolidate five levee boards into two. Though I agreed that a different levee board might be needed for the most complex urban-flood protection in the nation, I was at odds with the rationale that the business community was using in the email:

> It's obvious that the destruction of thousands of our homes, our possessions and the loss of over a thousand lives has failed to even phase [sic] the misguided actions of the Orleans Levee Board. Clearly the Levee Board failed in its basic mission of inspecting the levees and floodwalls while it focused on secondary missions of building recreational islands in Lake Pontchartrain, casinos, lavish lunches after "drive-by inspections of the levees," water fountains and leasing spaces for sailboats and restaurants.[247]

These were the same unsupported facts that the committee chairs of Nagin's BNOBC had used. Again, I felt like a character in a B-rate movie, fighting against almost everyone. When I requested backup data, the response was always the same: "What's the matter with you? Are you saying you want corruption?" Acquaintances of mine who wore a "One Levee" lapel button only repeated the same sound bites when I asked for supporting data. An attorney colleague of mine observed that it looked to him like the Citizens for 1 movement was an activity to pass the time for wealthy, uptown women whose homes hadn't flooded.

On September 29, an Associated Press news article went coast to coast with this opening line: "Louisiana voters decide Saturday whether to overhaul the multitude of boards that oversee the New Orleans area's levees after the agencies were accused of cronyism, inefficiency and ineptitude in the wake of the 2005 hurricane."[248] The Louisiana constitution was being changed over accusations and allegations. And I seemed to be the lone person who thought that was strange. Not surprisingly, the legislation passed. The assets (e.g., marina and airport) once managed by the Orleans Levee Board were transferred to a separate board called the Non-Flood Protection Asset Management Authority.

"Finally, we're seeing the light of reform in our state,"[249] said a gloating State Senator Boasso, who could see his dreams of being governor shaping up rather well.

* * *

I could have fretted and mourned the terrible press that the people of New Orleans were getting. But I chose, as always, to look forward. And I continued my new life as a civic activist while Stanford adjusted to his second year of high school. My days were full of this new work of organizing events, writing press releases, and refining and redefining our tactics and strategies. Around this time, I realized that getting supporters together in one room was energizing and uplifting for them. So, I arranged what would be Levees.org's first "general meeting." We needed a guest speaker, so I extended an invitation to Dr. van Heerden. We explained that we were just a little group of volunteers, but we could offer him a small fee for his time. The good doctor accepted and appreciated the speaker's fee, which could at least pay for gas. With the help of Levees.org advisor Deirdre Boling, we secured space from the First Presbyterian Church at the corner of South Claiborne and Jefferson Avenues. We invited our supporters to join us at seven thirty on the evening on October 3. We invited the press too.

The meeting was remarkably well attended. We had set up chairs for about fifty people, and we had to get more chairs because more kept arriving. I started the meeting by thanking everyone for coming, and then I introduced Stanford after asking him to stand up.

"I want you all to meet the most important person of all—my son Stanford—who designed the Levees.org website and logo," I said proudly. "Without Stanford, there would be no Levees.org."

I waited patiently for the applause to die down, and then lifted the microphone to proceed with the meeting. But I stopped in surprise because the applause was still going on. In fact, it was getting louder. And people were standing up! I put down the mike and turned to watch this unfold before me. Then I looked across the pulpit at Stanford, who was shifting from foot to foot, smiling and looking a little embarrassed. I had never felt prouder.

When everyone got seated, I introduced Dr. van Heerden, who spoke with eloquence and conviction of the sheer magnitude of the Army Corps' levee-building mistakes. During van Heerden's talk, I learned why the mainly residential region of eastern New Orleans had flooded. He explained that, for the areas along the GIWW (Gulf Intracoastal Waterway), levees failed because they were filled in many places with erodible material, like sand, instead of good, thick Louisiana clay.

How was it possible, I wondered, for sand to be in the levees? What sort of quality-control personnel could have allowed that?

After thanking Dr. van Heerden, I took the microphone and explained that we had decided to open chapters in other states—starting with Florida and California—to help bring the issue of safe levees to everyone. Then, I spent a few minutes giving our supporters a tool. I explained how much power they could wield by correcting wrong information in the media. I encouraged them to send an LTE of two hundred words or less, politely describing what was erroneous, and then to provide the right data. One letter, sent the same day as dozens of others, had a powerful effect on news editors.

I closed by inviting everyone at the meeting to sign up for committees, most importantly the new letter-writing committee.

The meeting was a success, and the icing on the cake was a flattering editorial by Baton Rouge's *The Advocate.*

* * *

On November 6, 2006, Democrats took control of both the House and the Senate. A few days later, I read in the *Times-Picayune* that Representative Waxman, a member of the Government Reform Committee, would investigate the Bush administration, looking for possible wrongdoing in the federal government's response to the 2005 flood. This gave me an idea! Perhaps it was time to push for a truly independent commission to investigate the levee failures rather than just the IPET, which was basically the Army Corps investigating itself.

I called Vince to tell him about my idea, and he instantly saw the possibilities. A commission would be consistent with Levees.org's mission of education about the true cause of the 2005 flood. He also agreed that the climate in Congress would be different now. We decided to hold a conference call with Senator Landrieu to see if she would consider sponsoring such a commission. I called her office and set up a phone conference with the senior senator for two o'clock on Thursday, November 30.

Meanwhile, with Vince's assistance, I penned an opinion piece on why a commission was needed and sent it off to the *Times-Picayune*, expecting yet again to be rebuffed. But, this time, the media outlet recognized our existence for the very first time, accepted my piece, and published it. My op-ed article closed with this observation: "Congress authorized $738 million in 1965 to the US Army Corps of Engineers to build a flood protection system for Greater New Orleans. Had the Corps done their job, the hurricane's toll on Louisiana would have been some lost shingles, wet ankles and soggy carpets. And Michael Brown might still have his job at FEMA."[250]

On the appointed day of our phone conference with Senator Landrieu, Vince drove to my home where I have a landline with a General Electric speakerphone. Vince explained the etiquette to me, such as how to address the senator and what we would say. For a few minutes, we rehearsed our lines. Promptly at 2:00 p.m., a staffer rang our phone number and asked us to please wait a couple of minutes for the senator.

After the introductions, Vince spoke first: "At this time, Senator, I would like to draw your attention to something very important for the people of New Orleans," he said. "And that is the need for a truly independent and comprehensive investigation of the levee failures and flooding during the August 2005 hurricane."

After spending a few more minutes explaining that the IPET was managed by the Army Corps, an obvious conflict of interest, the senator appeared to agree. She then asked us to build support for such a commission, still unnamed, and to call her for a second meeting after we had built that support. And just like that, the conversation was over. Vince appeared to understand what Senator Landrieu meant by "building support." It meant getting the leaders of other organizations to sign on in support of the commission. But not just any leaders. We needed to get the support of leaders who represented the city: civic, political, environmental, and neighborhood.

We spent the month of December 2006 contacting group leaders to build an initial foundation of support on which to stand and had little trouble getting endorsements. Right away, we signed on the League of Women Voters of New Orleans, the Citizens Road Home Action Team, the Lake Pontchartrain Basin Foundation, and Save Our Wetlands, Inc. We got endorsements from the New Orleans City Council and St. Bernard Parish Council as well as from Congressman Bobby Jindal, Congressman Zach Wamp, State Sen. Julie Quinn, representing District 6, and State Senator Walter Boasso, representing

District 1. We received broad support from neighborhood leaders from all across the region. It was a tall order, and we did it. After all, every single one of us had suffered due to the failure of federally built floodwalls.

* * *

On Christmas Day, Stanford showed me something that he had just discovered about the Army Corps. Using free software called StatCounter, he had found out that the Army Corps was watching us. Army Corps personnel were using their computers to log onto the Levees.org website. Over the course of the preceding twelve months, the Army Corps had paid 416 visits to our site and viewed an average of three pages per visit.[251] I did not have a problem with that and was actually a little flattered. It cost the American taxpayer little for Army Corps personnel to be viewing our website. I thanked Stanford and turned my attention to 2007.

Chapter

6

Figuring Out the Allies

The year 2007 rang out with criticisms from faceless folks who did not see the need for an independent levee study. On January 4, I sat down to my desktop computer and found an unsigned email from jr7402@bellsouth.net, telling me that "if governor blanco and ray nagin had done their job locally to begin with, we would not need 8-29 investigation. also if those thousands of people would have simply left when ordered, the death toll would have been much much less."[252]

By now, I had developed a ritual for emails like this one:

- I would study it and see if it contained anything of value. (In fact, this email did. It touched upon the lack of a plan for those without a car, credit cards, road experience, or family living outside of New Orleans. This email taught me that, going forward, I should be careful to discuss only the flooding destruction of New Orleans. Why the levees broke was the mission of Levees.org.)
- I would save the email because it could be valuable later.

* * *

Vince Pasquantonio and I had decided to host a press conference announcing our levee-commission idea on February 5 in the historic Lower Ninth Ward. The night before, I was sitting with friends, watching Super Bowl XLI, when my cell phone rang. It was Aaron Viles and I could tell right away that something was wrong. Viles was communications director for a local environmental group called Gulf Restoration Network. Viles and I had met each other recently to chat about our similar goals and about being allies. But he had bad news for me that night.

Weeks before getting Viles's phone call and understanding how critical it was to have a broad and diverse support base, Vince and I decided to approach his friend Julie Allen ("Miss Julie"), whose home in the Lower Ninth Ward flooded when the Industrial Canal breached. We wanted to use her teetering, damaged home as a backdrop for our press conference. Miss Julie was happy to oblige.

During my call with Viles, he told me that he had heard from two civic activists in the historic neighborhood about our press conference the following morning. The leaders—Linda Jackson, president of the Lower Ninth Ward Homeowners Association, and Patricia Jones, director of recovery for the Lower 9th Ward Neighborhood Empowerment Network Association—objected to our using Miss Julie's home for our press conference. Viles suggested that I call both Ms. Jones and Ms. Jackson to explain what we were doing and to extend a personal invitation to each.

Understanding my mistake, I immediately called Ms. Jackson and then Ms. Jones. I extended a personal invitation and offered each an opportunity to say a few words. They both accepted. Later, I heaved a huge sigh of relief. It was a close call. I should have realized that it was important to invite leading members of the community, even though I had Miss Julie's permission to use her private property. After all, this was their neighborhood, not mine.

* * *

February 5 dawned sunny and cold. The press event was scheduled for eleven in the morning, so I had time to review my notes. We had decided to name the bipartisan study the 8/29 Commission with the goal of examining the flood-protection design and decision-making since 1965 when the New Orleans metropolitan flood protection was federalized. Such a study was needed, we would say, because the Army Corps' levee investigation was a conflict of interest in that the agency was managing an investigation of its own work. In contrast, we would point out, the 8/29 Commission would be unbiased and nonpartisan. It would be federally authorized and financed because Congress was not compelled to recognize the results of independent studies. In closing, we would say that it could be done concurrently while flood-protection plans were moving forward and would not delay ongoing work.

I put on a Japanese-inspired blue jacket and plain black pants and drove to Miss Julie's house in the Lower Ninth Ward. As I turned left onto Tennessee Street and drove toward 5001 N. Rocheblave Street, I was struck, yet again, by how barren and bleak the neighborhood looked. I arrived at Miss Julie's house about fifteen minutes before the press conference.

I jumped up and down to stay warm as I waited for the press and invited guests to arrive. Both Ms. Jackson and Ms. Jones arrived as well as several council members from St. Bernard Parish. As for the press, we had success! Every television station showed up. One of them even brought a ladder mechanism to capture aerial views of the event.

And so, against the backdrop of Miss Julie's uninhabitable home, Vince, H.J., and I began our call for a federally financed study of the levee and floodwall failures that had doomed New Orleans and St. Bernard Parish seventeen months earlier. I invited each of the elected officials and the community leaders to say a few words, and then closed the event. Afterward, every reporter from every television station approached me for an on-camera response to a specific question.

* * *

When I returned home, there was an email from John "Spud" McConnell, inviting me to appear on his show on WWL (AM) radio at 5:00 p.m. I offered to drive to Spud's studio, but he assured me that it would be fine if I called into the studio from my home. At the appointed time, I dialed in and the interview began. After Spud asked me why Levees.org was calling for the 8/29 Commission and I had responded, he started taking calls from listeners.

The first listener was named David, who wasted no time in making his point: "You have no credibility! You're being paid by Joe Bruno and the lawyers for the lawsuit. You have no credibility!" he hollered.

I calmly replied that his statement was not true. I stated that if I were to see Joe Bruno on the street, I would walk right past him because I didn't know him.

"You have no credibility! Good luck with the lawsuit!" he yelled and hung up.

My mood improved around six o'clock when I got an excited call from Vince. A story by Cain Burdeau with the Associated Press had been featured in the *Washington Post* and 180 other media outlets all across the nation.[253] The well-written, balanced article also included a fabulous photograph. The reporter had obtained a comment from

Wayne Stroupe of the Mississippi Valley Division of the Army Corps. Apparently, when Burdeau requested comments from the Army Corps in New Orleans, he had been sent up the chain of command. Stroup said that an 8/29 Commission wasn't needed because the IPET included "150 national experts from more than 50 organizations and was reviewed by two independent panels."[254] He described our demand as similar to a 9/11 Commission in an apparent effort to disparage us. (The 9/11 Commission was experiencing unpopularity at that time.) Nonetheless, this was national press—the second time this had happened for us. It was a splendid ending for our press conference and a good start in building support.

* * *

With Vince's guidance, we continued to build broad support for the 8/29 Commission. Separately, to avoid repetition and burnout, Vince and I reached out to the leaders of neighborhood groups all over the city and requested a few minutes to speak at their upcoming neighborhood meetings. After the 2005 flood, residents' attendance at these meetings was at an all-time high, which worked to our benefit. Gone was the belief that one could manage one's life and well-being alone. The devastation of the 2005 flood had shown us just how fragile we—and our connections—were and how much we needed them to survive. For several months, Vince and I attended neighborhood meetings in St. Bernard Parish, Lake Bullard, the Lower Ninth Ward, the Upper Ninth Ward, Bywater, Fillmore Gardens, Broadmoor, Lakeview, and many others.

A week later (February 12, 2007), Dr. van Heerden offered his name in support. Vince contacted the chair of the local chapter of the National Association for the Advancement of Colored People (NAACP) while I contacted the chair of the Association of Community Organizations for Reform Now, which at that time had its national headquarters in New Orleans. I reached out to the

National Council of Jewish Women. And we visited local chapters of environmental groups, including the Sierra Club and the National Wildlife Federation. All these organizations provided a written endorsement for the 8/29 Commission.[255]

* * *

In April 2007, the State of Mississippi got 72.5 percent of $400 million that Congress appropriated to help FEMA develop alternatives to trailers and mobile homes. Louisiana got 20 percent with the rest going to Alabama and Texas.[256] Senator Landrieu found this inexplicable. The continued belief in Washington, DC, was that the people of Mississippi had been clobbered by Mother Nature and that the people of New Orleans and Louisiana were flooded due to their own ineptitude.

In a testament to those assumptions held by members of Congress, I observed that survivors of the flooding in Mississippi reported an entirely different story than survivors from Louisiana. For example, Bay St. Louis resident and architect Ed Wikoff recalls, "Although I could have done without the whole experience, in the end I think that Mississippi's handling of the reconstruction process was great."[257] In contrast, in Louisiana, homeowners struggling in Louisiana reported that losing all their belongings was less painful than dealing with the reconstruction process in their state.[258]

* * *

In this dark world, I found solace in a *New York Times* editorial, which chided Congress for not waiving the Stafford Act for regions that had flooded on August 29, 2005. The Stafford Act provides incentives to states by requiring them to ante up twenty-five cents for every seventy-five cents provided by the federal government. However, in cases of extreme disaster, the White House can "waive" the Stafford Act, which, since 1985, it has done thirty-two times, including for

Florida after Hurricane Andrew in August 1992 and for New York after Sept. 11, 2001.[259]

"This inaction is particularly surprising," wrote the editors of the national media outlet, "given that such a large proportion of the damage can be attributed to the failure of the federal levees that were supposed to protect the New Orleans area."[260]

* * *

On January 12, while on my daily internet search for anything related to the 2005 flood, I ran across a blog called *FoodMusicJustice.* I was intrigued by one particular post titled, "It was the levees, stupid." The post began with this statement: "A majority of Americans believe two things about New Orleans that simply aren't true. 1) A hurricane destroyed the city. 2) New Orleanians are responsible for what happened to them because they recklessly lived in harm's way."[261]

I wanted to see the rest of the post, but the link didn't work! I wrote to the blog's owner—a media expert named Ken McCarthy—who wrote back right away and provided a new link.[262] I dived into the post and found out that Ken and I were on the same page regarding the rampant myths and how they were hurting New Orleans's recovery. I wrote back, thanked him, and complimented his work. Lightning-quick Ken responded, explaining that he had worked in internet promotions since 1993. "I'd be very happy to help you in any way I can. I'm also happy to help with advice on internet strategy," he wrote. "There are many free and inexpensive things you can be doing to get your message out."[263]

And that is how we met. Ken's early advice and guidance would eventually catapult Levees.org to its place as an influential grassroots group with high name recognition.

Ken lived in New York State but had plans to visit New Orleans for a week in February, and we scheduled a meeting at my home.

At my kitchen table, in less than an hour, Ken laid out a strategy for using the internet and video. I scribbled furiously on a notepad, taking down everything. But it was his advanced advice that caught my attention. He had suggested creating a public service announcement (PSA) video and bringing it to the local television stations.

At that time, the local stations were required by law to run a minimum number of PSAs submitted by nonprofits at no charge. The challenge was that the video needed to be of broadcast quality. I hired Francis James of Perception Films. Francis was a beloved adjunct visual-arts teacher at Isidore Newman School with Hollywood experience. While Francis started working on concepts for the PSA, I focused on attracting a celebrity. Ken had told me that a celebrity is not of critical importance, but if I could get one, that would help.

I had just met a local celebrity on January 4, at a panel discussion at Loyola University. This event was planned in conjunction with the arrival of more than 450 law students from all over the country, who were in New Orleans during their winter break doing volunteer recovery work. There were several panelists; journalist John Barry, R. King Milling, leader of the advocacy group America's Wetland, and Harry Shearer, actor, comedian, and voice actor on the popular television show *The Simpsons.* Shearer was a part-time New Orleans resident.

I made quite a spectacle of myself at the panel discussion by challenging both Milling and Barry on what I saw as untruths in their statements. I didn't have to challenge Shearer, so after the event, I approached him to tell him that. Shearer told me that he was first alerted about the Army Corps' responsibility for the 2005 flood when he saw Levees.org's yard signs saying, "Hold the Corps Accountable." He said that they seemed to be all over the city and that propelled him to research the issue further.

After my meeting with Ken McCarthy, I contacted Shearer and invited him to appear in a PSA.

"We cannot compensate you," I told him, "but the PSA will be shot and edited by a professional. The entire experience will be all pro."

"Sure," he said, giving me the name and email address of his assistant to make the arrangements.

And, just like that, I had a celebrity face and voice for our first PSA!

I had to miss the video shoot due to an annual family reunion in Colorado. I asked Vince if he could meet Shearer on February 23 at 706 Phosphor Avenue in Metairie. Shearer was in very capable hands with Francis, who had experience working with names like Tom Cruise, Heath Ledger, George Clooney, and Billy Bob Thornton. After the shoot, Vince wrote to me saying that everything had gone "swimmingly."

* * *

On March 14 we held a press conference to release the first in our planned series of PSAs, starting with the Harry Shearer PSA.[264] In our press release, we stated that copies of the PSA on Betacam SP tapes would be available at the press conference and also online on our website as high-quality QuickTime downloads. WWL-TV Channel 4 and WGNO-TV Channel 26 showed up for the press conference, and the other two major network television stations (WDSU-TV Channel 6 and WVUE FOX 8 Channel 26) reported on the story in their nightly newscasts. *Gambit*—a popular free weekly newspaper—ran a story, and so did the *Times-Picayune* news bureau, which had finally stopped ignoring Levees.org. Decorated *Times-Picayune* reporter Mark Schleifstein's story appeared on page 3 of Section A. It was press galore.

The next day (March 15), the founder of Capitol Hill Broadcasting Network offered to upload our PSA free of charge since we were a nonprofit organization providing valuable information to a national audience. Over the next few weeks, other news stations such as WLAE-TV called me for a copy of the tape to play on their stations.

* * *

Meanwhile, Vince and I felt that we had collected enough support for the 8/29 Commission. It was time to meet with Sen. Landrieu and find out if she was as good as her word. Through her local office, we scheduled a short meeting while she was in New Orleans. We were told that Sen. Landrieu had a few minutes to spare just before a meeting at the Hyatt Regency on April 20.

At the appointed time—3:00 p.m.—Vince and I waited nervously with our long list of supporters in our hands. Vince prepped me and explained that, when the senator entered the room, we should immediately stand up. Just before three o'clock, the senator arrived and, after rising like our chairs were on fire, we showed her our impressive list of support.

"I'll do it," she said after looking for perhaps five seconds at our list. "Contact my 'leg director' at my DC office." And off the senator went to her next meeting. All the hard work and all those meetings had paid off. My mind whirled with thoughts on how to tell our supporters that, yet again, their time supporting Levees.org was well invested. Now that we had the support of the senior senator, we needed also to get Sen. Vitter's support. That would be a challenge, but what worked to our advantage was his disdain for the Army Corps, specifically its slowness.

Vince and I scheduled a conference call with Sen. Vitter's advisor, Garret Graves. During the call, Graves said that if we would craft the language for the 8/29 Commission, he would be happy to review it. We used the investigation into the Space Shuttle Challenger disaster in 1986 as a model. According to Dr. Bea, the expense should be about $5,000,000—less than one-tenth of 1 percent of the cost to rebuild the flood protection system. Vince and I could not imagine anyone being against such an investigation.

* * *

While Vince devoted a significant amount of time to the 8/29 Commission, I—buoyed by the success of the Harry Shearer PSA—decided to enlist the help of superstar actor John Goodman. Goodman lived part time in New Orleans. His home had been flooded badly due to the 17th Street Canal floodwall breach. My key to reaching Goodman was his daughter Molly. She and my son Stanford had been in the same class at Isidore Newman School since pre-kindergarten. I asked Stanford to speak to Molly at school and find out if her dad would like to say a few sentences in a video. A few days later Stanford came home from school and handed me a phone number on a piece of paper.

"Molly says that her dad would like to do it," said Stanford. "She says to just call him anytime."

I was bit nervous about calling him. Through all the years of parent-teacher association meetings and school events, I had never spoken to Mr. Goodman. He was usually away working, and, when he did attend school events, he talked to the parents of the girls. After all, he had a daughter!

I wrote out what I would say before picking up the phone and dialing the number. When he answered, I explained that we were creating a PSA in a professional studio.

"We'll mike you up, you can say your lines, and you will be finished in fifteen minutes," I explained.

He said that he would be traveling for the next two weeks. But, as soon as he got back, we would schedule something.

"Would you like me to contact an assistant to schedule the shoot?" I asked.

"No, you can call me."

And, just like that, I had lined up one of arguably the most famous actors in the United States to appear in a PSA! Three weeks later (Friday, April 6) at ten thirty in the morning, Goodman met me and Francis at 706 Phosphor Avenue in Metairie. I hung around and made myself as inconspicuous as possible. But I wanted to be sure that Goodman said the script as I had written it: "Hi. This is my town—New Orleans, Louisiana—where we were protected by an Army Corps federal levee. With levees in every state of this nation, and over 120 possibly vulnerable to this kind of disastrous failure, we all have a lot at stake. This land is our land. Please go to www.levees.org and join us. Don't we all deserve levees that work?"

He was brilliant, and I could not have been happier. I was also very conscious of the fact that Goodman was offering his face and voice at no charge, and I did not want to keep him even one minute longer than necessary. But I did ask for a photo!

* * *

For the John Goodman PSA, we decided to employ a different look from the Harry Shearer video. Using the mosaic that Stanford had built with photos of flooded homes, Francis created a cascading shower of photos, which all fell perfectly into place. On May 9, precisely at noon, we posted the John Goodman PSA to YouTube and followed Ken's formula to the letter. This meant that also, precisely at noon, we sent an email to our supporters announcing the second celebrity PSA. The email asked our supporters to go to YouTube and view the thirty-second video. We explained that it was critical to click the link to the video within twenty-four hours—the window in which YouTube collects viewing data that determines which videos are placed on the homepage of YouTube where it can potentially be viewed by millions.

The results were remarkable. In twenty-four hours, the video garnered nine honors, including #9 top rated that week in News & Politics.[265] Ken McCarthy boosted the numbers further by sending the video to his significant database, which added another five thousand clicks. This action put us at page number one in that category. This is remarkable success for a fledgling nonprofit organization of our size.

WDSU-TV Channel 6 emailed me and requested a copy of the PSA to put on the air.[266] The station also invited me to do a prerecording for its noon show the next day (Friday, May 11). Then, on that same day, a Yahoo! producer contacted me with the subject heading "showing PSA on Yahoo! front page?" The producer wrote that Yahoo! was interested in featuring the Goodman PSA on their homepage, which at that time was the most trafficked site on the internet.[267] This would give Levees.org enormous visibility. Straight away, I contacted Francis and asked him to send Yahoo! the links they needed. Then, all we could do was wait.

* * *

Two days later was Mother's Day (May 13). I told my family that I didn't want to do anything special, wanting only some peace and quiet—a chance, even for just one day, to relax and recharge. I played tennis that morning and spent the remainder of the day catching up on emails. For dinner, Steve cooked one of my favorite dishes—garlic chicken à la Mosca.

After dinner, I was back at my desktop computer, finishing up with the constant flow of email. At about 8:15 p.m., I received an email with the subject heading "are you crazy?" Then another email appeared with the subject heading, "you wear your politics on your sleeve." Then a third said, "are you all nuts?" It suddenly struck me: perhaps the video of John Goodman had just gone up on Yahoo!'s website. I logged onto the site and, sure enough, on the upper right was our John Goodman PSA! The Goodman video was up for about three hours that night. And in that space of a few hours, it got 40,000 views.

With regard to the hate mail, I followed my ritual. I reviewed each one, looking for new information, operating under the assumption that there could be as yet a piece of data that was being withheld from me and would come as a broadsided surprise. When I found nothing new, I filed them away and did not look at them again. There was no room in my life—and certainly not on Mother's Day—to feel sorry for myself.

The next day, I received a call from a female staffer with the Army Corps Public Affairs Office. She wanted to know if I had paid John Goodman for his appearance in our PSA. I replied that no, that would be absurd. Goodman is a superstar.

A few days later, the Army Corps quietly hired a public relations firm called Outreach Process Partners (OPP). This small firm, which had an office in Maryland and in New Orleans, was paid one million dollars up front.[268] Over the next three years, the firm would receive another $3,613,998.00.[269]

* * *

Three days after the Goodman video appeared on Yahoo! (May 16), Vince and I were on a plane to Washington, DC, to drum up support on Capitol Hill for the 8/29 Commission. Vince, who had attended Georgetown University, knew how to navigate the city's train network from the airport to Capitol Hill. The system was efficient and inexpensive.

That same morning, Cheron Brylski issued a press release: "This week, representatives of Levees.org are on the Hill talking to members of Congress about the importance of passing the 8/29 Commission, an unbiased investigation of the failure of the federally built levees."[270]

Our directive from Senator Landrieu was to get support from four senators outside Louisiana in the Senate Committee on EPW (Environment and Public Works). We visited them and also Senator Feingold. We also met with Graves in Senator Vitter's office. He sat down with us and, line by line, went meticulously through our text for the 8/29 Commission.[271] Graves raised some valid points, and we agreed to make the edits in return for Senator Vitter's agreement to cosponsor the bill.

The next day (May 17), I sent an email to the entire membership of Levees.org, alerting them of our accomplishments. I shared that Senator Landrieu had filed S.2826 and promised us that she would make the 8/29 Commission her number one priority in the conferencing of the WRDA bill.

A few days later, (May 21), I received an angry phone call from Senator Feingold's legislative director. She said that she had seen the bill's summary and told me that it was not what she expected. Perplexed, I looked up the summary of the bill from my desktop computer and found this text:

> 8/29 Investigation Team Act—Establishes a bipartisan 8/29 Investigation Team to examine: (1) the events beginning on August 29, 2005, with respect to the failure of the flood protection system in response to Hurricanes Katrina and Rita; and (2) each

> flood control and restoration project that has been carried out since that date in the region in which those events occurred.[272]

I explained that I understood her concern and agreed that this summary did not look right at all. I promised to call Senator Landrieu's office and reiterate what we wanted: a team to review all the findings and recommendations of all studies conducted in the aftermath of the levee failures in Louisiana on or after August 29, 2005. A month later, the proper wording for the bill itself was added to the congressional website, but the confusing summary language remained, even after I pointed out that it was misleading.

* * *

Dr. Seed continued to fret over what he was seeing. He had attended two major conferences dealing with disaster and forensics, which, naturally, had last-minute sessions hastily scheduled (after the programs had been set) to deal with the issues related to the 2005 flood. One of those sessions was organized and chaired by Dr. Mlakar with the Army Corps, and the other session was with Larry Roth, executive deputy director with the ASCE. Dr. Seed was dismayed that neither his own independent investigation team (the Berkeley team) nor Dr. van Heerden's team (Team Louisiana) were invited to participate in either conference.[273]

* * *

On June 1, my attention was pulled abruptly away from Capitol Hill and celebrity PSAs. The long-awaited ERP report had just gotten released, nearly two years after the 2005 flood. H.J.—normally the careful one who prefers to ponder something before deciding on it—called me.

"We have to issue a statement immediately...today," he said. "The ERP report itself is very good, but the introduction to the report is awful."

It was true. The eighty-page report, with helpful photos and graphics, was quite good and written for a layperson to understand. But the accompanying press release contained information that 1) was not in the report, 2) conflicted with the report, and 3) minimized the Army Corps' involvement in the catastrophe.[274]

For example, the press release stated, "Even without breaching, the hurricane's rainfall and surge overtopping would have caused extensive and severe flooding—and the worst loss of life and property loss ever experienced in New Orleans."[275]

In fact, the ERP report stated something quite different. It said that, had the levees and pump stations not failed, "far less property loss would have occurred and nearly two-thirds of deaths could have been avoided."[276]

After H.J. and I prepared and released our view of the ASCE's press release and sent it to our supporters, I received a forwarded email from Dr. Bea. It was written by Gordon Boutwell, a member of the ERP and a fine civil engineer, who wrote, "ASCE HQ changed our stallion into a gelding."[277]

While the members of the Berkeley team, the ERP, and Team Louisiana discussed their collective response, H.J. and I pondered even further actions we could take as an advocacy group. That's when I remembered that I had met a civil engineer at Senator Landrieu's office a week before while Vince and I were in DC. His name was Stephan Butler, and he had insisted that I call him if I needed anything. I dug up Butler's business card and emailed him. I told him that I wanted to know how much money the Army Corps had paid the ASCE to convene the ERP.

Two weeks later (June 21), Butler forward an email from Larry Roth. Roth confirmed that the Army Corps paid the ASCE about $900,000. He also wrote, "I will be pleased to speak with Sandy Rosenthal at any time. From what I understand, we have much in common. My contact information is below."[278] That afternoon, I dialed the Reston, Virginia, number and Roth answered. After I introduced myself, Roth said, "Hold on. Let me close the door so we won't be interrupted."

I waited for what seemed longer than necessary to shut a door. Finally, he picked up the phone and explained that he would be putting me on the speaker phone and asked if I would mind. I complied. (Months later, I would learn that the Berkeley team members had been subjected to this identical thing.)

The conversation began with me pointing out that there were inconsistencies between the report and the press release. We questioned the wisdom of the Army Corps directly paying the organization responsible for peer reviewing the IPET report.

The next five minutes were a steady monologue from Roth, who spoke about the integrity of the people who served on both the IPET and the ERP. Throughout his speech, his tone of voice steadily rose. He ended with this:

"These upstanding people don't have so much as a traffic ticket, unlike you!" he said.

"I pay my traffic tickets," I replied quietly.

"Sure, you do," he replied with a sneer.

I changed the subject and pointed out that, if a building full of people crashed to the ground, you wouldn't blame the janitor. You would look for the architect, the contractor, and the engineer. In the case of the 2005 flood, the Army Corps was all three.

"The Army Corps of Engineers is NOT solely responsible for the flooding!" Roth fairly shouted.

Understanding that the conversation was going nowhere, I thanked Roth for his time and hung up. The traffic-ticket reference was telling. Truth be told, just six weeks earlier I had negotiated a moving-traffic violation down to a seat-belt violation. Apparently, Roth had done a background check and knew my deep, dark secret. It was just a shade unnerving—but just a shade. That day, for the second time since the 2005 flood, I told myself that it would take a lot to intimidate me. It would take more than a headless bird or a background check to scare me away from what I believed needed to be done.

I called H.J. and told him about Roth going ballistic on the phone. H.J. told me that he would call Roth too. An hour later, H.J. called me back to say he had received the exact same treatment—a steady stream of angry defensiveness. Curiosity got the best of me, and I tried a search on the internet to find a video or an audio of Roth.

I did find Larry Roth. He had given a PowerPoint presentation at Auburn University in Alabama eleven weeks earlier (April 5).[279] I listened to the entire eighty-minute video presentation.[280] While the PowerPoint slides and graphics were accurate enough, Roth's narration repeatedly made statements that drew attention away from the Army Corps and onto the storm, the city's geography, and, most of all, the Orleans Levee Board. At one point, he drew laughter from the audience while making fun of the Orleans Levee Board—a clear breach of ethics.[281]

Whoever crafted the accompanying script was skilled at slanting the material in one direction. Anyone not intimately familiar with New Orleans and the 2005 flood would conclude, after hearing Roth's narrative, that the Army Corps had done the best it could but were hamstrung by local officials.

It was now crystal clear to me that the engineering establishment was shielding the Army Corps from criticism.

What I would not know until later was that this was one among dozens of presentations that Roth would present from September 2006 to February 2008.[282] These presentations were essentially public relations efforts designed to protect the Army Corps' reputation among communities that it could reward with lucrative contracts. Congress awards hundreds of billions of dollars to the Army Corps for water projects. The Army Corps chooses who can approach this trough of work. With money comes power. But, as nauseating as this revelation was, I could not see that there was anything I could do about it beyond take notice.

* * *

A week later, Boutwell sent me a letter with further details on why the ASCE's June 1 press release was unsatisfactory. Boutwell closed with this statement: "Overall, I found the Press Release selective; that its parts contain the truth but not the whole truth. Its objective appears to be exoneration of the designers (Army Corps) from any responsibility. I am proud of ASCE for publishing the ERP report and the cited article. The Press Release makes me ashamed for my Society."[283]

In a phone conversation on June 21, I asked Boutwell if I could disseminate his letter. He responded no, because, according to him, "someone at the Army Corps called him and said that, if he didn't stop talking, the Corps would destroy his firm."

I was bitterly disappointed, but I understood. The Army Corps rewards civil-engineering companies that keep quiet and just do their job. But the Army Corps can also punish those who cross them. This dilemma brought harm to many, including Levees.org. For example, that same week, a tearful supporter asked me to remove her name from our mailing list because members of the Army Corps were patrons at her husband's restaurant for lunch at the Riverbend. They needed the Army Corps' business.

* * *

Our lives running an advocacy group continued to play out. All summer, we were called upon by local and national media for comments about the flooding. Only about half of our interviews made it into print or television, and we were often misquoted. But we did not judge ourselves by whether or not media outlets covered our efforts.

The most important thing that happened the summer of 2007 was receiving Jon Donley's invitation to host my own column. Donley was founder and editor-in-chief of NOLA.com, the online version of the *Times-Picayune*. Having my own column allowed me and Levees.org to take advantage of NOLA.com's significant search-engine power and further increase the reach of our message.

Each time I published a blog post to Levees.org, I made a copy and posted it on my NOLA.com column with a different title. But there was a second reason that having column privileges with NOLA.com would turn out to be important. Eighteen months later (December 10, 2008), these privileges would allow me to make a shocking discovery. But, for now, my NOLA.com column was just another way for me to increase the reach of Levees.org's message at no cost other than my time.

* * *

National news focused on how the Army Corps was rebuilding the levee system, but always failed to mention that the federal agency's floodwall building mistakes were the cause of the disaster. For example, in an August 26, 2007, story by Brian Schwaner with the Associated Press, a photograph caption stated, "…if the Army Corps of Engineers has its way…those in New Orleans should be better protected."[284]

This statement implies that, prior to the 2005 flood, the Army Corps did not get its way. It fueled the fairy tale.

* * *

On August 30, 2007, the day after the anniversary of the 2005 flood, H.J. and I attended a special closed-door meeting in the Tulane Conference Room, fifteenth floor, of the New Orleans Marriott Metairie at Lakeway. Larry Roth, ASCE executive deputy director, and Bill Marcuson, president of the ASCE, wanted to meet with me and other stakeholders in the New Orleans region.

When H.J. and I arrived, Marcuson walked toward me and, with pomp, said, "I am Bill Marcuson, president of the American Society of Civil Engineers."

I resisted the temptation to say, "I know who you are."

Marcuson introduced me to Casey Dingy, managing director of external affairs for the ASCE. We all sat at an enormous oval conference table and greeted Paul Kemp (Team Louisiana) and also Luke Ehrensing (the Berkeley team) and Gordon Boutwell (ERP), both civil engineers. H.J. and I sat directly across the table from Roth, the same man who had lambasted us both nine weeks earlier (June 21). But on this day, Roth—a bald, overweight man—was more peaceful. On either side of him sat Marcuson and Dingy.

The conversation began with what was foremost on everyone's mind: the inaccurate press release about the ERP. Roth, in an immediate show of contrition, admitted that, in writing the press release, he had made the biggest mistake of his career. "Can't we just let bygones be bygones?" he asked.

Ehrensing stuck his hands in his pockets, leaned back in his chair, and replied that he and others in the room had something else in mind. "A retraction," he said abruptly.

"We will consider it," replied Marcuson.

The next order of business was to view a PowerPoint presentation. I instantly recognized the very same show from Auburn University, only this time Roth's narrative was different. Roth did not draw out in excruciating detail the low-lying geography of city or the strength of the hurricane. He did not poke fun at the Orleans Levee Board.

At the conclusion of the presentation, I spoke first: "Mr. Roth, I have seen this PowerPoint before. You gave this at Auburn University this past April.[285] But, this time, you gave a completely different narrative."

Roth sat motionlessly and stared at me from across the table. I stared back with the calm that comes from having thoroughly done one's homework.

I continued: "Today, your narrative was balanced, but the narrative you gave at Auburn shifted blame away from the Army Corps and onto the people who live in New Orleans."

Roth suddenly tried to stand up. Quick as cats, the two men on either side of Roth grabbed him by the shoulder and pulled him back down.

"That's only what she believes," said Dingy softly to Roth, who sat slowly back into his chair.

The entire room of people watched in silence as this played out. Eventually, Ehrensing broke the spell with a technical engineering question about Roth's presentation, and the tension diffused. The meeting adjourned shortly afterward, and WWL-TV Channel 4's Lee Zurik entered the room to take some interviews. That night, congratulatory notes were sent among the attendants, excluding those with the ASCE. Ehrensing noted that Roth would be giving the PowerPoint presentation again at the Louisiana ASCE meeting that coming weekend, and he joked about "which PowerPoint we were going to hear this time."[286]

While H.J., Vince, and I felt the ASCE PowerPoint presentation was a collaboration with the Army Corps, we could not prove it. So, we did what we could, which was to use this presentation as further evidence that the nation needed a truly independent, bipartisan investigation.

I pored over the curricula vitae of all the participants of the IPET and found that 80 percent worked for either the Army Corps or its sister agency the ERDC. Furthermore, of the top three IPET cochairs, two worked for the Army Corps, and one of them was there for fifteen years (1986 to 2002). Of the twenty-three task-team leaders, six worked for the Army Corps and seven worked for the ERDC. We did not need more proof than that, so I put together a two-pager, explaining this.[287] In addition to our who's who, Vince and I decided to change the name of the 8/29 Commission to the 8/29 Investigation, to avoid touching a sore spot with some members of Congress who considered the 9/11 Commission controversial.

On September 26, I flew to DC and met with Senator Harry Reid (D-NE) about the need for the 8/29 Investigation and also to see Paul Rainwater, chief of staff for Senator Landrieu. I showed him the two-pager. Rainwater flipped through it, apparently quite interested, but said nothing. When he was done, I began my impassioned speech, describing the need for an investigation that the people of New Orleans could trust. Rainwater listened in silence. Not feeling that much was accomplished but understanding that I had to try, I returned home.

* * *

Later that month, at Ken McCarthy's urging, I requested a meeting with a political commenter and New Orleans resident, the well-known and influential James Carville, to discuss the need for the 8/29 Investigation. Carville and I agreed to meet at his home on State Street. I gave him the presentation that Vince and I had honed down to tight talking points. At the conclusion, Carville did not have any suggestions regarding the investigation, but he did have a comment.

"To do what your group did with no paid staff is a story all by itself," he said. For a few minutes, Carville talked to me about ways to make Levees.org a dependable, credible bank of levee data. Later, as he walked me to the door to say goodbye, he repeated his observation about what Levees.org had accomplished with so little. I thanked him and put the compliment out of my mind. There was too much work to do.

* * *

In mid-October 2007, Dr. Seed called me. After seeing that the August 30 meeting accomplished nothing except an apology behind closed doors from Roth, Dr. Seed had decided to file an ethics complaint with the ASCE. He wanted to share his letter with me, which he did. The forty-two-page letter laid out in stark detail the whole story of apparent collusion between the Army Corps (Mlakar) and a key ASCE staffer (Roth). Dr. Seed asked me not to share it until he had submitted it to the ASCE. He was forbidden from sharing his letter with the press, but I was free do with it as I pleased. For the time being, I elected to keep it close to the vest.

* * *

For months, I'd had a fantasy of creating a satirical video, spoofing the cozy relationship that I had witnessed—long before reading Dr. Seed's letter—between the ASCE and the Army Corps. In my mind's eye, I had seen a suitcase stuffed with money changing hands, accompanied by knowing winks among Army Corps employees. It was the time to make such a video, but I needed a professional to bring this idea to life.

I called Francis James. He accepted and suggested we make the video a school project at Isidore Newman School. Francis selected Lori Bush's US Government and Politics class. After getting a green light from the school's headmaster, Francis and I got straight to work. I wrote a script for the sixty-second video, and Francis designed the scenes around the dialogue. He suggested that we use rubber tubes, rafts, and floaties as props. For three days, I called all my friends who had pools to gather these items while Francis found a briefcase, fake money, and other props. One of the students in the class suggested that the actors who would receive the briefcase wear "bling" or ostentatious jewelry. All our proverbial ducks were lined up for the video shoot.

On Thursday morning (October 5), we arranged for the entire video to be filmed in one morning. Then, during lunch, we had pizza delivered and invited eight teachers to sit at student desks and pretend to be students while the students play-acted as teachers. The effect was hilarious. Lights! Camera! Action! We wrapped up just as the bell rang for the first afternoon class.

One week later, we rolled out the video, which we called "Levee Spin 101," at a press conference and applied Ken McCarthy's tried-and-true formula to promote it.[288] Within thirty-six hours, the video got 16,000 views and fourteen honors on YouTube.[289] The Isidore Newman School teachers forwarded the video to their families and friends nationwide, urging them to view and rate the video. Four days later, the video had been viewed over 25,000 times, which, in 2007, was an extraordinary number for a nonprofit organization of Levees.org's size.

* * *

Saturday, November 10, dawned sunny and cold, but windless. I put on my tennis clothes and headed to Bissonet–Maned Downs Country Club to play a singles match with Debbie Cobb. Even during busy weeks like this, I refused to give up my precious exercise. After I returned, showered, and ate some lunch, I got back to my desktop computer and the ever-present incoming emails.

I noticed an email with the subject heading, "Cease and Desist Letter," which was sent from an attorney representing the ASCE. The headmaster of Newman School was also copied. I opened the attachment and discovered that the ASCE didn't like the "Levee Spin 101" video and were ordering me to take it down from YouTube. After a lengthy description of the history of the ASCE, the letter ended with this: "…should you ignore this letter and continue to disseminate this defamatory material, please be advised that ASCE intends to take appropriate legal action to protect its interests."[290]

This was a technical way of saying, "If you don't stop, we will sue you."

My initial reaction to this letter was that we must be on the right track! I thought it was quite remarkable that an in-house counsel attorney felt fine about harassing a fledgling grassroots group and a bunch of high-school kids. This was likely the first time in ASCE's history that the society had to deal with negative press of this magnitude.

I wrote to ASCE's in-house counsel Tom Smith and told him that we would be in touch on Monday, November 12. Then I disseminated the cease-and-desist letter to Cheron Brylski (my pro bono publicist), my advisors, the Levees.org board, and, of course, Dr. Seed. All responded within minutes, including Paul Harrison, an attorney for the Environmental Defense Fund and an ally in Washington, DC.

Harrison remarked, "I'd expect that this would be an appealing pro bono cause for one of the big guys."[291]

* * *

On Monday, November 12, my phone rang off the hook. Ms. Bush called to tell me that one of her students at Isidore Newman School had asked her in tears, "Ms. Bush, am I going to get sued?" Email after email flowed in asking question after question.

Smith wrote back to me, saying, "I received your email but continue to see that the video remains on YouTube, which now shows over 29,000 views. Please let me know when to expect further communication or action."[292]

What were we to do? My next-door neighbor called me. An attorney in a small family firm, he offered to represent me at no charge. He was confident our circulating "Levee Spin 101" on YouTube was protected free speech.

Twenty minutes later, another attorney called me, this time from Adams and Reese, which in 2007 employed about 275 lawyers and was one of the largest law firms in the southeast. He introduced himself as Martin Stern. In a calm voice, he told me that, while he and his firm were willing to represent me, there could be a problem. Should the ASCE file suit in Reston, Virginia, which was the location of ASCE's headquarters, then Adams and Reese could not represent me. There lay the challenge.

After conversing all day with other members of the Levees.org board and with Stern, we decided to take the intermediary step of taking down the video while we examined our options.

I sent an email to ASCE's counsel, saying, "We are a small grassroots organization and do not have the resources to go to court, and thus have to acquiesce to your intimidation tactics. We'll take the video down when our webmaster, my seventeen-year-old son…comes home from an out-of-town school track meet tonight."[293]

I hated to back down, even just a little, but we needed an attorney to fight an attorney, and we didn't yet have one in Virginia—not yet.

* * *

Also on Monday (November 12), Dr. Seed submitted his ethics complaint to the president of the ASCE. Dr. Seed's letter was long. It meandered here and there and often repeated whole passages. Instead of writing a forty-two-page letter, I dearly wished that Dr. Seed had written a ten-page letter with thirty-two pages of appendices. Be that as it may, the letter made it clear that Dr. Seed had witnessed an early, systematic, intentional plan by the Army Corps and the Department of Defense to:

- Hide the Army Corps' culpability in the New Orleans flooding.
- Limit, control, and discredit the two independent investigations.
- Limit the scope of the official IPET investigation.
- Delay the release of the final results until the public's attention turned elsewhere.

I offered Dr. Seed's letter as an exclusive to Grunwald with *Time* magazine. If he chose not to accept the exclusive on the story, I asked him to let me know by noon the next day (November 13). I was getting more savvy and understood that I needed to provide a deadline.

That night, when Stanford got home from his track meet, he changed the setting for "Levee Spin 101" from public to private, making it accessible only to those with "viewing privileges."

* * *

The next day (November 13), the ASCE's president took the unusual step of issuing a letter to every member of the Louisiana ASCE, vigorously defending itself. Also on that day, while I was being interviewed by John ("Spud") McConnell on his WWL (AM) radio show (*The Spud Show*) about the video, Rudy Vorkapic, editor-in-chief of the satirical newspaper *The New Orleans Levee*, called into the show. On the air, Vorkapic told me that, if I would send him the video, he—in the name of journalism—would upload "Levee Spin 101" onto his website, where anyone who wanted to see it could do so.

"Let ASCE try to sue me!" he said. "Make my day!"

* * *

Sometime before noon, Grunwald called me from *TIME* to say that his boss had not given the okay to accept the exclusive. I hung up and called John Schwartz with the *New York Times*. As with Grunwald, I offered him an exclusive and asked, if he chose not to accept it, to let me know by noon the next day (November 14).

Late that night, I received a surprise phone call from a producer with the *Times-Picayune*. Despite a history of ignoring the work of Levees.org, they wanted to post our spoof to the NOLA.com website. At 12:09 a.m. on November 14, the *Times-Picayune* did just that, which brought the video even more attention and visibility.[294] The ASCE's plan had backfired. Meanwhile, the vicious comments posted to the video were a sign that we were hitting the ASCE and the Army Corps where it hurt.

For example, this comment was posted by someone with the username "swain":

> Shame on Sandy for using her kid for a cheap publicity stunt. Obviously her group has NO EVIDENCE WHATSOEVER and so she put her own child up to making a video of SLANDER and INNUENDO. You can't just call Asce liars and think they aren't going to defend themselves.[295]

This was one of dozens of comments that I read and put out of my mind in order to focus on the work at hand.

* * *

Noon on Wednesday (November 14) came and went, and I still had not heard from Schwartz about the exclusive I offered him. When I called to remind him, he launched into a long explanation about why he could not accept my offer. Allowing my irritation to show, I cut him off, saying all I needed was a yes or a no. Another day wasted. I then called Cain Burdeau with the Associated Press and gave him the customary twenty-four-hour period to consider the exclusive.

As I was managing the dissemination of Dr. Seed's letter to the press, the ASCE's cease-and-desist letter continued to draw the attention of television, radio, and print outlets who wanted to interview me. If the pressures of this nonstop media attention were not enough, I was also hosting a houseful of relatives! My sister Melissa and my two brothers Mike and Rusty were in town to attend an awards ceremony for Stanford, who was being honored as the first ever youth recipient of a prestigious philanthropy award. Amidst all of this, I needed to let Stanford know that I was so proud of him. So, while the scenario was unfolding better than in our wildest dreams, the physical toll was tremendous.

This became apparent on Thursday (November 15), the morning of the awards luncheon. I awoke early and got ready for a seven o'clock show with Spud McConnell on WWL (AM) radio. I was tired from being pulled in so many directions, but I felt that I should do the preplanned interview. It began with McConnell asking me about the video, why we created it, and why we took it down from the Levees. org channel on YouTube. Then he began to take callers.

The first caller was named David. It may have been the same David who dialed into McConnell's show nine months earlier who hollered that I had no credibility. In a taunting voice, he asked why the investigators should work for free. I responded that this was a case of conflict of interest. When he started to ask a second question, I interrupted him.

McConnell pressed a button in his studio, which muted my voice, and yelled, "Hey, hey, hey! Let the guy talk, for Pete's sake!"

It was a huge mistake. I sounded impatient and unreasonable. Later that day, comments were posted to the *Times-Picayune* story about how rude I had been on the Spud McConnell show. I learned another valuable lesson: never lose my cool. I also realized that there may be times when I should forego an interview.

* * *

Just as the McConnell interview ended, the phone rang. It was Martin Stern with Adams and Reese, and he had some good news for me. In his calm voice, he assured me that, by creating and posting our video, we had indeed exercised a fully protected right to free speech. We agreed to have a phone conference in one week.

What a morning—and it was only eight thirty! Remembering that I had a houseful of guests, who were all from the Boston area, I brewed a big pot of tea. Folks from Boston like to drink strong, dark Lipton tea with lots of milk and sugar. I pulled myself away from my emails to spend time with my sister and two brothers before getting dressed for the awards luncheon.

* * *

On our way to the Sheraton, we picked up Stanford from school, which had given him permission to miss two midday classes. The awards ceremony was a large, noisy event with many familiar faces. Just before it was time to be seated, a tiny, attractive, and very loud woman with red hair fought her way through the crowd toward me.

She introduced herself as Dr. Laura Badeaux, executive director of the Louisiana Center for Women in Government. After congratulating me for Stanford's honor, she told me that she also had congratulations for me: her organization had just voted to induct me into their Hall of Fame for my work on the 8/29 Investigation. I thanked her, and gently let her know that this was Stanford's big day and that I needed to get to our table. When I found our table, I turned my focus on Stanford and my family, who had traveled far to be there.

In the car, after the ceremony, holding Stanford's large, crystal, pyramid-shaped award for Outstanding Youth in Philanthropy in my lap, my cell phone rang. It was Burdeau with the Associated Press, calling to tell me that his boss had not approved the exclusive.

I hung up, sighed, and called Schleifstein with the *Times-Picayune*, who accepted the exclusive on the spot. He was fourth on my list because I had tried to get national coverage first. But, even if the coverage for Dr. Seed's letter was local, I was confident that Schleifstein would do an excellent job. I had been reading his articles for two and a half years, and I had a lot of respect for him, though not for his bosses.

* * *

The next day, I drove my siblings to the airport and apologized profusely for having spent so little time with them, promising to make it up to them. That night (November 16), I sent an email to our growing number of supporters who deserved to know what had transpired over the "Levee Spin 101" video. I explained that, after receiving the threat of a lawsuit, we realized that we had neither the personnel nor the resources to take on a legal battle with a large, powerful organization such as the ASCE. I told them that, when the video was taken down, it had nearly 30,000 views and had earned nine honors.[296] I took great delight in telling our supporters that, in the following three days, the video was viewed another 14,000 times and garnered twenty honors, including number-one-most-viewed video in the category Non Profit.[297] An international version of YouTube—Alternative Channel—offered us free space from then on to post our videos.

* * *

Late the following Monday night (November 19, the first day of the Thanksgiving holiday week), Schleifstein's article about Dr. Seed's letter went live.[298] It painted a picture of collusion, control, and delay.[299] It also corroborated everything in our "banned" satirical video.

What a wonderful, stressful, amazing, and difficult seven days! And as my family and I slid into the quieter period of the Thanksgiving holiday, the feeling that stood out most for me was pride. I was so proud of my son for his giving spirit. And I was proud of the Isidore Newman School students who were now famous faces after all the news coverage.

* * *

For the next several days, the phone stopped ringing due to the holiday, but the emails continued to pour in. I let some of them pile up in order to take a much-needed reprieve and spend time with my family. Two days in a row, at 5:30 p.m., I participated in step-aerobics classes. It always surprised me how any kind of exercise could magically transform big, impossible problems into small, manageable ones. On the day before Thanksgiving, I drove to the airport at two o'clock to fetch my son Mark, who had flown in from Denver. I drove back again at 10:20 p.m. to meet my daughter Aliisa, who had flown in from New York City. We were all together again.

Thanksgiving went just like it did every year for the previous twenty years. My husband Steve was the cook, and I was the maid. Steve cooked fourteen dishes from scratch: turkey, gravy, oyster dressing, traditional stuffing, mashed potatoes, sweet potato casserole, mirliton (or chayote, a light-green, pear-shaped squash) stuffed with shrimp, broccoli casserole, steamed cauliflower, homemade cranberry sauce, three pies (apple, pecan, and pumpkin), and a chocolate cake. I entered the kitchen every thirty minutes to clean bowls, utensils, blenders, knives, pots, and pans. And we all loved every minute of it. This year, like every year, when my sister-in-law Leslie Jacobs's family and my mother-in-law Sandra Rosenthal and her husband Rogene Buchholz arrived at four thirty, we never breathed a word about how hard we had been working all day. What mattered was being together, enjoying Thanksgiving dinner with a nice bottle of wine.

* * *

At midday on Thanksgiving Day (November 22), I saw one of the most important emails I would ever receive during my leadership of Levees.org. The three-day-old email had a subject heading, "the video fracas," and it was from Samantha Everett, a California attorney who let me know that her firm, Cooley Godward Kronish, had offices in Reston, Virginia—the same location as the ASCE headquarters.

Ms. Everett had received my email, forwarded from a Levees.org supporter, where I had explained that we didn't have the resources to fight a lawsuit in Reston, Virginia. She wrote, "I'll do everything I can to find you excellent, free legal assistance."[300]

Even though it was Thanksgiving Day, I responded immediately: "I am very interested in your offer. Though we have received a half dozen offers of pro bono service, we had to acquiesce to the ASCE because if the ASCE sued, it would likely be in Reston, VA. Clearly, your offer changes everything."[301]

I forwarded Ms. Everett's note to Martin Stern, who emailed back early the next morning (November 23). He said that Adams and Reese would be happy to partner with Cooley Godward Kronish and that he would call Ms. Everett. In subsequent three-way phone conversations with the attorneys, we determined that the ASCE's threat of a lawsuit had fit the legal definition of a strategic lawsuit against public participation (SLAPP). In other words, the ASCE's threat of a lawsuit was intended to intimidate and silence us by threatening to burden us with the cost of a legal defense. Cooley Godward Kronish, in partnership with Adams and Reese, was prepared to defend us by relying on anti-SLAPP legislation, which was designed to protect citizens like me from such abuse of power. The two attorneys asked me to take some time to think about it since nothing was ever a sure thing.

Feeling that I needed some outside guidance, I reached out to Adam Babich with the Tulane Environmental Law Clinic, whose mission was to assist nonprofit organizations like Levees.org. On Tuesday, December 11, I drove to the Tulane campus and met with the bespectacled man, who had thick black hair touched with gray. He was already familiar with our case, and he advised me that, since we had gotten so much press, we had accomplished our goal and that the best course of action was to not repost the video.

I replied without hesitation. "If the press you speak of was national press, like the *New York Times*, I might have followed that counsel. But this is the *Times-Picayune*! This is not what I consider significant press!"

I thanked Babich for his time and felt, for the first time, utterly certain that we would publicly and ceremoniously thumb our noses at the ASCE's threat of a lawsuit and repost the video. All I needed was a time and place. I called Wendy Carlton, a Levees.org supporter who lived near the breach of the 17th Street Canal floodwall. Her family had just rebuilt their flooded home in the Lakeview neighborhood and the house was still bereft of furniture, making it a good venue for television cameras with plenty of space for invited guests.

Next, I called Martin Stern from Adams and Reese and told him that it was a go. Levees.org would repost "Levee Spin 101" on YouTube at a press conference on Friday, December 14, at 10:30 a.m.

The next day (December 12,), Ms. Everett sent a letter to Tom Smith, the ASCE's in-house counsel. The letter told Smith that his November 10 letter to Levees.org, copying the Isidore Newman School, was an attempt to bully a small nonprofit organization out of exercising its First Amendment rights and that Levees.org would repost the video to YouTube in two days. The short letter closed with this statement: "We believe that the anti-SLAPP statute of Louisiana will apply to this case. Please be aware that should you file suit against Levees.org, we will vigorously pursue a judgment against ASCE for all of the fees and costs incurred by this Firm and Adams and Reese, LLP."[302]

On the same day (December 12), I issued a press release to local and national media announcing our plans. Christi Lu, a staffer with the Army Corps' New Orleans District Public Affairs Office, intercepted my press release and forwarded it to the top of the chain: the Army Corps headquarters in Washington, DC.[303]

In response, Steven Wright, also with the New Orleans office, issued an urgent message to his fellow administrators the next morning (December 13) in the DC office (Eugene Pawlik, Suzanne Fournier, and Gregory Bishop): "This is a national Corps issue and requires a USACE-level response to query. Should you get media calls on this subject please refer callers to Mr. Gene Pawlik 202-761-7690."[304]

Meanwhile, Pawlik forwarded the email to Joan Buhrman, an ASCE spokesperson in Reston, and asked her, "Do you have any information that we might be able to share with our leaders if they ask?"[305]

Pawlik's "leaders" would be none other than Donald Rumsfeld, the US secretary of defense.[306]

* * *

On Friday, December 14, reporters and camera operators for every television station in town showed up in addition to the *Times-Picayune* and WWL (AM) radio. Flanked by attorneys from Adams and Reese—Martin Stern and Lauren Delery—I began the press conference.

I started with a recap of why Levees.org partnered with the Isidore Newman School to create the video and then discussed the threat of a lawsuit by the ASCE to silence us. The legislative director from Levees.org, Vince Pasquantonio, spoke for a few minutes on the need for a truly independent levee investigation. Then, with camera crew in tow, I walked upstairs to Ms. Carlton's home office, sat down at her desktop computer, and, with a click of a mouse, changed the status of "Levee Spin 101" from private to public.

I stood up, turned to the cameras, and declared, "Today, Levees.org showed New Orleans, Louisiana, and the whole nation that the American Society of Civil Engineers shall not bully, shall not intimidate a little grassroots group and a bunch of high-school kids out of exercising our First Amendment rights."

That night, according to Harry Shearer, the story of our thumbing our noses at the powerful ASCE was the subject of every news-media outlet in town, including print, television, and radio. WWL-TV Channel 4 did its first ever news story about our work. And, once again, the students from Ms. Bush's US Government and Politics course at Isidore Newman School were local celebrities.

* * *

On Friday afternoon at 4:25 p.m. (December 14), after the press conference, Pawlik, the administrator at the Army Corps' DC Public Affairs office, wrote Ms. Buhrman, the ASCE spokesperson in Reston, asking, "Any insight from ASCE on this?"[307]

Ms. Buhrman replied, "Per our conversation, here is a copy of the ASCE statement. Feel free to call my cell if you need anything else."[308]

This statement was attributed to David G. Mongan, the new president of the ASCE: "The American Society of Civil Engineers sent the letter to protect our reputation. Levees.org took the video down and we considered this issue resolved. Since the video has already been widely reposted by other organizations, moving forward, we feel our time and expertise are best utilized working to help protect the residents of New Orleans from future storms and flooding."[309]

In response to the public relations crisis created by Dr. Seed's letter and Levees.org's satirical video, the ASCE confirmed that it had launched two ethics investigations. One of them was an internal probe, led by staffer Rich Hovey, to look into Dr. Seed's claims in his forty-two-page letter. The second was external, led by retired Congressman Sherwood Boehlert, and was set to examine the activities spotlighted in our spoof: the methods of the ASCE when it received millions from organizations that selected the ASCE for peer reviews.

* **

A few days before Christmas, Matt Faust—a young man from St. Bernard Parish—contacted me. He had spent hundreds of hours creating a hauntingly beautiful video masterpiece from family photos that his grandmother had remembered to pack before she evacuated ahead of the 2005 hurricane. Before making the video, Matt had consulted with his LSU professor about using copyrighted music by Jon Brion in the video. The professor, it turned out, had given him bad advice, telling Matt that, so long as he didn't charge anything for the video, it was allowed. This was not true, and, before long, Matt got a nasty letter from Brion's attorney, threatening a lawsuit if he continued to disseminate the six-minute video.

Perhaps it was confidence over our freshly won victory over the "Levee Spin 101" video.

Perhaps it was anger over the activities of powerful people with deep pockets who use attorneys to make themselves feel more important.

Perhaps it was just the Christmas spirit of giving.

Despite the risk, I told Matt that I would put the video, titled "home," on the Levees.org website, and Brion's lawyer could go ahead and sue me. The video was a masterfully illustrated story about how a young man's flooded childhood home was torn down by accident in the post-flood environment of confusion and chaos. I asked all of Levees.org's supporters to view it. A total of nine hundred people visited the website on December 21 and 22. And a total of 4,698 people spent an average of 3:37 minutes on our website after viewing the video until August 12, 2008.[310]

Each time I felt frustrated during the next three years about the slow pace of progress or felt tired with the difficulty of each step, I watched "home." The video gave me perspective and showed me that my worries and concerns were minor compared to this family from Chalmette in St. Bernard Parish. It also gave me peace because, as hard as it was to lose all of one's possessions, we still had each other. Often Matt's video made me cry, but it was a good kind of crying—the kind that makes me feel a little better than I did before.

Despite the chest-thumping, Brion's attorney never sent Levees.org a cease-and-desist letter. And, on December 21, the Army Corps' New Orleans District Public Affairs Office circulated my email about "home," noting that the Army Corps was not mentioned in it.[311]

Chapter

7

Foiling the Ruse

The year 2008 began dreadfully.

In January, US District Court Judge Stanwood R. Duval Jr. dismissed the class-action lawsuit against the Army Corps. The plaintiffs, represented by Joseph Bruno, wanted compensation for the failure of the 17th Street and London Avenue floodwalls. Judge Duval found the Army Corps guilty of "gross negligence" but was unable to award damages to the plaintiffs because the Army Corps was protected from liability by the Flood Control Act of 1928.[312] In other words, the Army Corps was guilty but suffered no financial consequences.

Duval was dismayed over having to make his ruling, as demonstrated by his conclusion in the forty-eight-page opinion: "This story—fifty years in the making—is heart-wrenching. Millions of dollars were squandered in building a levee system with respect to these outfall canals."[313]

Adam Nossiter, a reporter for the *New York Times*, invited me to comment on the ruling and closed the article with my remarks, word for word: "Clearly Judge Duval is frustrated by what he had to do.

It's outrageous these levees were fragile. He and I agree the corps was responsible for the failure of the levees. It's a positive thing that Judge Duval outlined all those things in his statements."[314]

In the following weeks, follow-up news articles were dreadful because they painted a picture of greedy survivors of the 2005 flood who would not get the cash they hoped for. I was surprised by this in light of the reasons for which Judge Duval dismissed the case and his denouncement of the Army Corps' negligent work.

This situation prompted me to pick up the phone and call one of Levees.org's advisors, Mark Davis, who by this time was head of the Tulane University's Institute for Water Resources Law & Policy. I wanted to know if statements made in an opinion by a federal judge could be put forth as fact.

"Absolutely," replied Mark. And, in his droll manner of speaking, he added, "Perhaps not cosmically, but, in America, yes, statements in a federal judge's final opinion can be considered vetted fact."

Mark closed the conversation with this pithy observation. "The American people, unless there is a financial award, tend to dismiss a case as having no merit."

* * *

After the ruling, things did not improve for the people of New Orleans.

In Washington, DC, Congress voted to require the State of Louisiana to pay $1.8 billion of the cost to fix levees and floodwalls and pay it back in just three years. This was unfathomable since the direct and proximate cause of the 2005 flooding was the failure of the Army Corps' floodwalls. But it went even deeper. The Army Corps knew over one hundred years ago that building levees along the Mississippi River to benefit cargo navigation and commerce would starve Louisiana's coastal wetlands of replenishing soil.[315]

As observed by Gary Rivlin, "The sediment the river would otherwise be depositing at the mouth of the Mississippi was now ending up on the bottom of the Gulf of Mexico."[316] The loser was Greater New Orleans and south Louisiana because the marshy wetlands of the coast were a natural buffer against storm surge, which accompanies a strong hurricane. This damage eventually combined with the flooding's proximate cause: the Army Corps' defective floodwalls, which breached in the heart of New Orleans several feet below the design specs. Why wasn't Congress listening?

* * *

On February 27, Michael Chertoff, secretary of Homeland Security, told a group of reporters that other people besides the Army Corps should take the blame for the flooding of New Orleans and that more money should have been spent on maintenance."[317] A short time later, *Popular Mechanics* ran an interview with Eric Halpin, the Army Corps' flood reconstruction chief, who implied, in response to a softball question, that levee failure in New Orleans was due to poor maintenance.[318] Meanwhile, there was still no evidence that any of the levee breaches in New Orleans were due to lax maintenance. I used these off-base comments as rally calls. In response to both, I wrote to our supporters and urged them to "click here" to write to their members of Congress and demand the 8/29 Investigation.

"Don't we all deserve levees that work?" I asked, repeating the question in the PSA from John Goodman, Levees.org's spokesperson. I built a similar campaign in response to Congress's decision to force Louisiana to pay $1.8 billion in just three years.

* * *

It was a glum, dark time and I saw just how little top officials in federal government understood the 2005 flood. I searched for ways to give Levees.org supporters reasons to feel encouraged. It was at this time that I found out that I was being honored along with three impressive Louisiana women.

Each year, the Louisiana Center for Women in Government honored four residents in the State of Louisiana for their significant achievements or statewide contributions. In my case, I was selected for my role in mobilizing support for the 8/29 Investigation legislation. The Hall of Fame event provided heavy press for its annual event, but I felt torn. On the one hand, I understood that this honor was a positive thing for the mission of Levees.org; on the other hand, how could I talk about it without appearing to brag? Feeling lost, I decided to ask Ken McCarthy how to handle this dilemma.

Ken, as usual, knew what to do. And, in typical McCarthy-style, rather than explain it, he responded by writing a draft email for me to edit and send to Levees.org's supporters:

> As a testament to the success we've achieved as a group in focusing national and local attention on making sure the levee system in New Orleans is rebuilt correctly, the Louisiana Center for Women in Government has inducted me into their Hall of Fame. I'll be accepting this honor on behalf of all of the members of Levees.org. Thanks to everyone who has not only stood up for New Orleans, but also for communities around the country who are at risk because of substandard levee construction and maintenance. We still have a long way to go, but recognition of our efforts, as signified by this award, shows our voices are being heard.[319]

Ken showed me how to tackle the challenge of doing something that was at its heart self-promotion, but which, if handled right, could create positive energy for the public good. The email helped Levees.org's members see that their help was generating results and their time was well invested.

* * *

The mission of Levees.org was educating others that the flooding of Greater New Orleans in August 2005 was a man-made disaster due to the Army Corps' mistakes. Levees.org's first message was "Hold the Corps Accountable." Starting in January 2007, the mantra became "Demand the 8/29 Investigation." Each time we saw evidence of corruption or collusion on the part of the Army Corps, we demanded the 8/29 Investigation, which would cost five million dollars. Maintaining this consistent demand gave us an appearance of doing our research and demanding something reasonable. Ultimately, our push for the 8/29 Investigation died a quiet death. I never witnessed Senator Landrieu, the bill's sponsor, taking any steps of her own to garner further congressional support, and I will never know the reason why. Nonetheless, it was a powerful rally call and a splendid builder of our database. Not one moment of pushing for an independent, bipartisan investigation was a moment wasted.

* * *

During the fourteen years following the 2005 flood, I found that the most important discoveries would spring from unexpected places. This happened in early March 2008 when we received a response from the Army Corps to our very first request under the FOIA (Freedom of Information Act), which we had filed four months earlier.

When we released "Levee Spin 101"—our satirical video spoofing the Army Corps' relationship with the ASCE—we decided to file a request for records under the FOIA. Such requests might sound like a formidable undertaking, but they are not. Requesting documents under the FOIA requires little more than writing on a letterhead who you are and what you want, and a willingness to pay for the records if the request is huge. We asked for the formal documentation of the grant awarded by the Army Corps to the ASCE to perform the ERP. We submitted the FOIA on Friday, November 9, 2007. The very next morning, we received the cease-and-desist letter from the ASCE, so we didn't give the FOIA request another thought, having many fish to fry.

We had already moved onto other goals when the thick packet of material arrived. Not terribly interested, I thumbed through the pages to find out exactly how much money the Army Corps had paid the ASCE. According to the documents, the total was $1,150,000. At least we knew that the number stated in the spoof (nearly a million) was indeed accurate.

I sent the FOIA response to H.J. for his review, and he found something that I had missed! On page three, the grant document described how the ERP would be done in four phases:

- Phase 1 was research and analysis on the performance of the levees and floodwalls.
- Phase 2 was provision of information on the current system to prevent future flooding.
- Phase 3 was provision of information to evaluate alternative approaches to flood protection.
- Phase 4 was transfer of information and knowledge gained to a broader audience within the Army Corps and its consultancy community.

It was Phase 4 that was important. H.J. explained that "consultancy community" refers to the independent civil engineers that the Army Corps hires nationally to do 90 percent of its work. This could also include college and university engineering students, like those in the Auburn University video I had found online. The meaning was crystal clear. The Army Corps had paid the ASCE to go around the country, giving dozens of presentations that were designed to steer blame for the flooding away from the Army Corps (and the engineering community) and onto the people of New Orleans.

H.J. and I reached out to Dr. Bea (the Berkeley team cochair), asking if he agreed with our assessment: Phase 4 described direction and payment from the Army Corps to the ASCE (i.e., Larry Roth) to travel the nation and give more than thirty presentations from February through October 2007. Dr. Bea wrote back and replied in the affirmative.[320]

We now had proof that the PowerPoint presentation that Larry Roth gave to the Auburn students, and to others, was paid for by a grant from the Army Corps. And it was apparent, in Roth's narration that the intent was to shift blame away from the Army Corps and onto local officials. This was scandalous. This was evidence that the Army Corps—under our very noses—had paid the ASCE to improve its image after the 2005 flood, a flood that caught the attention of the entire world. And they called it "education."

We moved quickly, intending to ride the wave we created a few months earlier when we gave the ASCE "the bird" and reposted our video "Levee Spin 101" portraying the ASCE's cozy relationship with the Army Corps on YouTube. H.J. and I decided to hold a press conference to announce the finding.

I called Levees.org supporter Arnold Eugene, an African American resident from eastern New Orleans. I asked if I could visit him to see if his home might work as a press venue. We had already held a press conference in lily-white Lakeview. I felt that it was time for a similar venue in a black neighborhood. Eugene's home, in what was locally known as New Orleans East, had flooded in August 2005 mainly due to multiple breaches in the levees along the GIWW. 80

percent of the neighborhood of primarily middle- to upper-class black homeowners was flooded by more than six feet. Almost half the region sustained more than eight feet of water.[321]

The next day, I drove to Mayo Boulevard, just off I-10. As I turned left off the service road and onto the boulevard, it was clear that this was once a beautiful and stately neighborhood. The homes were all large and stood on once-grassy yards. But now the grass was killed by briny floodwaters, as were most of the trees. Eugene's recently rebuilt home was large and mostly empty of furniture, making it perfect for a press conference.

We set the date for Thursday, March 27. We would demonstrate that the ASCE's PowerPoint presentations, ostensibly protecting the Army Corps' reputation, were funded by the federal government.

At 9:35 a.m. on March 26 (the day before the press conference), Kathy Gibbs from the Army Corps' New Orleans District Public Affairs Office contacted the Washington, DC, office, alerting them that the New Orleans office would request their comment about our press conference.[322]

Four minutes later, Suzanne Fournier, the administrator at the DC office of the Army Corps, responded by instructing two DC Public Affairs officers to "draft a corporate response that could be used for media and for commanders in casual conversation."[323]

The news about our press conference in New Orleans—and the information it would reveal—had been kicked way upstairs.

That same morning, Wayne Stroupe with ERDC Public Affairs in Vicksburg, Mississippi, contacted Denise Frederick, the in-house counsel for the New Orleans District Army Corps of Engineers Office. Stroupe told her, "Media announcement below from Levees. org outlines press conference where they will make some serious misleading allegations. Have been asked for input by HQ USACE

Public Affairs office for talking points and messages for use by Army Corps. Will begin working on those."[324]

At 11:22 a.m. on March 27, just minutes before Levees.org's press conference began, Wade Habshey in the New Orleans District Public Affairs Office wrote to Rene Poche in the same office. He reported that Sarah Mclaughlin and Lee Mueller, two contractors in the Army Corps' office, were stationed at the press conference, were prepared to take notes, and would call him.[325]

Before the press conference started, Audie Cornish, a reporter with NPR, called me. She apologized that she could not make the presser but hoped to get a phone interview with me at two thirty. In keeping with my rule to protect myself from being overtasked, I passed this on to Vince.

At 11:30 a.m., our press conference began with good media attendance. All four of the local television stations were there as well as WWL (AM) Radio; Baton Rouge's newspaper, *The Advocate*, and *Popular Mechanics* magazine. We had promised video news releases (VNRs), which would explain how PowerPoint presentations hosted by the ASCE in 2007 were federally funded affairs crafted to mislead the engineering community and the American public on the Army Corps' role in the 2005 flood.

In the VNR, we first showed that the PowerPoint presentations contained at least ten falsehoods, four significant omissions, and numerous misrepresentations. We then supported our points with the data we found in the response to our request under the FOIA. In closing, we reminded the viewer of two things. First, members of the ASCE were forbidden from making false or exaggerated statements. Second, members of the ASCE were required to disclose when they are paid to make statements on behalf of a stakeholder.

In addition to the media attendance, there were also about two dozen Levees.org supporters present. After the event, H.J. and I took the opportunity to speak with each supporter. Then, in the normal fashion, H.J., Vince, and I were offered many interviews in the following days, including by Garland Robinette with WWL (AM) Radio and David Winkler-Schmit with *Gambit*. Perhaps best of all, on May 25, Dennis Woltering with WWL-TV Channel 4 invited me to his Sunday morning show for a twenty-minute, prerecorded, televised interview. This was the first time that WWL-TV Channel 4 had treated me like the expert I had become rather than just a rabble-rouser. What makes this notable is that, at that time, WWL-TV Channel 4, a local affiliate of CBS, was viewed by more than the other three television stations combined.

Day by day, press conference by press conference, rally call by rally call, we forged our way, leaving a record that we had been there. Levees.org shouted out loud its unwillingness to accept a lie that harmed people by its very telling. And all the while, the Army Corps watched our every move—not just in the New Orleans District but also in Division Headquarters in Vicksburg, Mississippi, and often in Washington, DC. It is clear that, for every radio show or interview, Stroup in the Army Corps' Mississippi Valley Division would issue an alert that a representative of Levees.org would be appearing in the media.[326]

A week later (April 2), the president of the Louisiana chapter of the ASCE sent an email to its nearly two thousand members, bemoaning our revelation of the facts: "We deem it unfortunate that a critic has chosen to concentrate their efforts on accusations against ASCE, rather than focus on the positive aspects of how they can foster and assist in positive results."[327] It contained the usual arguments (e.g., that we live and work here and must move forward). The ASCE didn't deny our statements; they just didn't like them. I tucked the email away and continued to drive the mission of Levees.org.

* * *

We began to sense that our work could be made easier by making the issue of floodwall failures relevant to people everywhere. It is normal to be unconcerned with something that does not affect one's own income, well-being, family, or neighborhood. As Harry Shearer had suggested to me in January 2006, it was critical that Levees.org portray levee safety as a national challenge. But how? The answer came, as many do, from an unexpected place.

On June 17, 2008, Mark Levitan, director of the LSU Hurricane Center, forwarded a press release for a planned news briefing in Washington, DC. It was titled, "Levee Protection: Working with the Geology and Environment to Build Resiliency" and was hosted by several hydrology organizations and the ASCE.[328]

Something stood out to me: the opening sentence of the news release said that the United States has thousands of miles of levee systems and that 43 percent of the US population lives in counties protected by levees.[329] 43 percent is a huge number, I thought. I made a few phone calls and found out that the person who might know what counties contained those levees was Gerry Galloway at the University of Maryland. He was scheduled to speak at the DC briefing.

On Monday June 23, a few days after Galloway's presentation, I managed to get him on the phone. I explained that I was fascinated to learn the 43-percent statistic. I explained that I wanted to know which counties contained those levees and how many people lived in each.

"I am sure that data are in a file cabinet somewhere at FEMA," he replied. "Why don't you ask FEMA?"

I thanked Galloway and hung up, thinking about the huge difference in stature between us. If I were to give a statement, I am quite certain the listeners would want me to support my statement with hard, indisputable data. But Galloway could get away with saying it "must be somewhere."

I filed another request for data—this time from FEMA. I felt that this data would be useful because, up to that point, we were having trouble getting levee safety to resonate as a national issue, even though levees seemed to be breaching all over the nation.

For example, in the middle of the night on January 6, 2008, a levee breached and flooded a town high in the mountains of Nevada. In the town of Fernley, 4,200 feet above sea level, an irrigation canal's earthen levee ruptured after heavy rains. The deluge flooded hundreds of homes and forced the rescue of dozens of people with helicopters and boats across a square mile of this desert town thirty miles east of Reno.[330] Six months later (June 12, 2008), serious flooding occurred in Cedar Rapids, Iowa, which is over 700 feet above sea level.[331] At about the same time, in Wisconsin Dells, 1,100 feet above sea level, a levee along Lake Delton, a popular tourism destination, breached and emptied the lake into the nearby Wisconsin River.[332]

These incidents did not seem to convince the American people that levee safety was a national issue. However, due to the June 19, 2008, public briefing notice we now had a cold, hard number—43 percent.[333] I filed the request under the FOIA and forgot about it.

* * *

The next day (June 1), Representative Melancon filed H.R.6526 in the US House.

* * *

The combination of the third anniversary of the 2005 flood and Hurricane Gustav brought enormous attention upon the mission of Levees.org. While there was plenty of good press during the "perfect storm," there was also some bad press. For example, an article from the *Boston Globe* stated, "Hurricane Katrina brought shame upon a fabled American city—and the nation. The infirm were abandoned. The lawless ran rampant. And vital government functions, most notably the protective network of levees, failed miserably."[334]

This press coverage was typical in that it failed to mention who was responsible for the failed levees. It also had the effect, whether or not intended, of blaming the residents of New Orleans for their misfortune. Up until now, I had forwarded news stories like this to a group of volunteers who enjoyed writing LTEs to correct wrong information and replace it with the right. They found this work cathartic, and it often educated the editors. Sometimes the reporters would write back and thank the volunteers.

But forwarding these news stories about the 2005 flood was starting to take up a good chunk of my time each day. Feeling the need to guard my precious energy, I decided to recruit someone for this job. From the pool of contacts who had been writing letters, I selected Melissa Smith for several reasons. She and her family lost the majority of their possessions when their Mid-City home flooded. I found Melissa's letters personal and filled with strong arguments. A New Orleans native, archivist, and professional writer, Melissa was a natural fit to lead the letter-writing team. We met at PJ's Coffee on Magazine Street uptown and crafted our strategy.

We decided that Levees.org needed a fresh pool of writers. The same faces had been writing LTEs since the 2005 flood. But we also needed to create a filter to catch Army Corps employees looking for intel while masquerading as Levees.org supporters. We decided to create a quiz in which interested people had to answer a few questions and provide basic information, including their full name, address, and phone number—necessary information in order to submit LTEs to media outlets. In other words, by asking for personal data, we were not ruining our chance to find bona fide quality letter writers. I built a levee quiz, using the online tool SurveyMonkey. I found out how difficult it was to create a good multiple-choice quiz. It took me four times longer than I expected. I also tried to keep the quiz from being intimidating by not using the words "quiz" or "test," so I called it "the levee challenge."

* * *

On November 17, I sent out an e-blast, telling supporters that they could join an important team to fight the myths and misinformation that the Army Corps' public relations company was dispensing. We received replies from hundreds of people.

From there, it was easy to ask the responders to take the levee challenge, which required them to input personal data. Melissa called each responder to confirm that the phone numbers were legitimate. Sure enough, some of the numbers were "not in service" and were eliminated. One of the names that we caught in our dragnet was Rene Poche. I recognized his name because he worked for the Army Corps' New Orleans District Public Affairs Office. In this case, Poche used his personal email address, and, of course, he was eliminated.

With the help of Ken McCarthy, we held two live training sessions at a local community center for local letter writers. For out-of-town writers, Ken hired Lily Keber—a young, smart artist—to video record a session, which we then posted to YouTube on a private account. Only YouTube subscribers who had been "invited" could view the video, which held all our "trade secrets" on how to do a successful letter-writing campaign. For a campaign to work, there can be no evidence of orchestration by a single person or an entity. Each letter must be unique without the appearance of being coached. We called the group of writers the CAT team (Counter Action Team) with the goal of countering bad or misleading information in the media.

Ten days later (November 27), after Melissa had screened all the candidates, we put together a large, local team and two national teams of letter writers, whom we affectionately called "cats," totaling more than 150 people.

The levee challenge quiz had begun as a tool with which to vet letter writers and catch infiltrators, but every year we improved the quiz in some way and rereleased it to our supporters, challenging them to find out how much they knew. This is an example of how media tools can be dusted off and recycled later on. The CAT team still continues its valuable work today. In a word, what makes the CAT team so effective is that it's composed of real people with real personalities passionately engaged in a cause and donating their time. And we have fun too. About every nine months, Levees.org hosts a luncheon, usually at Lebanon's Cafe, to share stories, laugh, and do what New Orleanians do best: have a party!

* * *

After waiting for two years, H.J. and I got news about the ASCE on April 6, 2009. The society's internal investigation concluded that Larry Roth's press release had indeed contained misleading information, the most egregious being the claim that widespread death and destruction would have occurred even without floodwall failures.[335] While this was welcome news, no ethics charges were leveled at any of the executives whom Dr. Seed had cited.

We were disappointed. We had expected calls for resignation for the ASCE's Larry Roth and the Army Corps' Paul Mlakar. And we had hoped for the light to be shined on the Army Corps' attempts to hamper the independent analyses.

To draw attention to what we felt was a whitewashed investigation, we held a press conference on April 8, 2009, to explain the meaning of these results. For the event, we selected the Lakeview

Harbor Restaurant at 8550 Pontchartrain Blvd. in New Orleans. In our email to supporters inviting them to attend, we included this teaser: "Levees.org will also present never before disclosed information."[336] That information was Boutwell's email criticizing the ASCE's press release, which he had asked me verbally not to share because he feared that the Army Corps would retaliate and ruin his engineering business.[337] Boutwell had recently died of a heart attack, so I decided to release it, believing he would want me to do that.

On the morning of the press conference, the Army Corps New Orleans District Public Affairs Office circulated an intercepted email about our press conference. In it, Karen Collins remarked, "It just made CNN headline news."[338]

At the press event, I released Boutwell's email written to me, describing his low opinion of the ASCE's June 1, 2007, press release. I also included the reason that I kept it undercover for two years. In my mind, the fact that Boutwell had refused to allow me to disseminate his email spoke to the control that the Army Corps had over the civil engineers and the big media outlets. We concluded the event by using this report as further evidence of the need for independent, bipartisan investigation.[339]

Five months later, on September 9, the other ASCE investigation—the external one—wrapped up. Congressman Boehlert released his investigative report that looked at the way the society received funding for doing peer reviews. The report scolded the ASCE for accepting funding directly from the organizations whose work it peer-reviewed. The Boehlert report went on to make several recommendations, the most important being that funding for peer reviews over one million dollars should come from a separate source, like the National Institute of Standards and Technology. The report also recommended that ASCE headquarters should facilitate—but not control—peer-review teams, and that dissemination of information to the public and press not be under the tight controls that the Berkeley team experienced. The report concluded that the ASCE should draw up an ethics policy to eliminate questions of possible conflicts of interest.[340]

A few months later (late November), the ASCE did issue its new manual describing how peer reviews should be funded in the future to avoid actual and perceived conflicts of interest. Mark Schleifstein with the *Times-Picayune* called me for a comment on the story. While I felt that the new funding rules were a major coup, I remained unhappy about the rules pertaining to experts and the media. The ASCE continued to prohibit team members from speaking to the media without approval of the ASCE's communications department until after a final report was released. To me, this muzzling of experts would have the effect of keeping critical information from the media and the American public for months or even years.[341]

A true hero in Levees.org's efforts to this expose this scandalous cooperation between the ASCE and the Army Corps is Ms. Everett, the original attorney who reached out to me from the other side of the country to offer her firm's pro bono assistance. (In 2010, the law firm wrote a note to me, confirming that the law firm's engagement in the lawsuit was terminated. The letter closed with this: "Congratulations on your continued success raising awareness about an issue of vital importance to not only the residents of New Orleans, but to much of America."[342])

* * *

On a clear, chilly, Wednesday morning in December 2008, I found myself instant messaging on Gmail with Bruce Biles, the owner of a popular blog called the *New Orleans Ladder*. Bruce, who had a significant following, was engaged with Levees.org's quest to separate the real from the fairy tale about the 2005 flood. We had been in contact since August 2007 when he ordered three of Stanford's mosaic posters. On that morning, we were chatting about the spike in vitriolic comments posted online to NOLA.com and to my Levees.org blog. The online comments were meant mostly for me, but they also attacked my supporters and the people of New Orleans.

For example, I found this online comment from "stevonawlins":

> Your performance to date suggests that you have chosen Corps-bashing as your mission with the expectation that when you manage to persuade enough under-informed people that Katrina was a federal flood, then the rest of the nation will pour even more money into the sinkhole of New Orleans.[343]

I remarked to Bruce that the comments were not more venomous than usual, just more frequent.

"You could find out where those comments are coming from," he suggested. "You just need the commenter's IP address. There's a website you can go to and just type in the IP address."

I pondered Bruce's suggestion. Initially, I had no interest. I had gotten this far by intentionally not focusing on ugly online remarks. I preferred to treat them as evidence that I was on the right track. But Bruce continued to encourage me, and I decided to pursue his suggestion, especially since it was so easy: just a few keystrokes.

And so, on that Wednesday morning (December 10), I logged into StatCounter, a free service for amateur bloggers. I scrolled through the visitor paths, which display the servers used by people who were viewing the Levees.org website. To my mild surprise, there were "hits" from several dozen people whose host was listed—in a bright-red font—as amvn91h.mvn.usace.army.mil. The telltale part is in the middle: the portion that says "usace," which stands for US

Army Corps of Engineers. So it was clear that dozens of Army Corps folks were visiting the Levees.org website regularly. This was not a surprise to me. But wait!

When I looked closer, I noticed a familiar IP address: 155.76.159.253.

I remembered the back-end tools available to me since 2007 when NOLA.com's editor-in-chief Jon Donley offered me an unpaid column. I quickly logged into my NOLA.com dashboard and scrolled through the litany of recent vicious postings. My heart skipped a beat. I read and reread the IP addresses. I blinked hard, thinking that something might be wrong with my eyes. The IP address was the same: 155.76.159.253. I had just found drop-dead, unequivocal proof that the comments posted by "stevonawlins," and many others—living, breathing people, not robots—were coming from the Army Corps' New Orleans headquarters. And I would never have made this discovery without Bruce's encouragement.

On the very same day that I made the discovery (December 10), the deputy commander of the Army Corps' New Orleans District updated and published the Policy Statement Use of Government Communications Resources, including email and internet access. The coincidence is remarkable, and I can only leave it at that.[344]

It was eerie that members of the Army Corps (the organization tasked to protect us) were using government computers to attack us. This was not a general case of mean-spirited, anonymous commenting. This was a specific case of people in a position of public trust disguising their identities, pretending to be objective onlookers, and using NOLA.com's comments section to attack people. The attacks did not discuss the issue; rather, they focused on me personally with the goal of discrediting me. I understood—or, more accurately, I felt in my bones—that Levees.org and I were the target of an organized campaign.

At that time, I didn't know the words to describe this activity. Anonymous commenting was new and unregulated in 2008. "Social media," as we know it today, barely existed fourteen years ago.[345] This was the Wild West of the online world. In December 2008, any vicious, anonymous comment posted online at NOLA.com—including outright slander—was allowed to stand. In that era, I had not even heard of anything like what was happening to me and my group. I did not know yet that there was a term to describe this activity: astroturfing. But I did understand that this activity was reprehensible.

As observed by New Orleans attorney Justin Zitler, "Such behavior is grotesque misconduct, a sustained disinformation campaign aimed at irredeemably undermining the words and deeds of an engaged civic activist speaking out for the public good."[346]

I picked up the phone and dialed Lee Zurik at WWL-TV Channel 4. He had done a good story about our decision to repost the "Levee Spin 101" video to YouTube. But it turns out he had just left for a ten-day vacation with his family.

I then asked to speak to Dennis Woltering, who had invited me seven months earlier to appear on his Sunday talk show. I told Dennis what I had discovered and offered him an exclusive. He was interested but had to check with his boss. Minutes later, Dennis called back, saying that it was a go, but there was a catch. He was off that week and would get on the story first thing Monday morning. I was disappointed, but because WWL-TV Channel 4 had such a high viewership, I agreed.

As Dennis had promised, he called the Army Corps first thing the following Monday morning (December 15). Then Dennis emailed me, saying that Timothy Kurgan, a major at the Army Corps, would not respond until he saw proof.

I chuckled and thought, "You want to see proof? I'll show you proof!"

Five minutes later, I sent the damning proof to Dennis: screenshots of the incriminating evidence. One screenshot taken from my NOLA.com column showed that "stevonawlins" was using the IP address 155.76.159.253. The other screenshot, taken from StatCounter for the Levees.org website, showed that the IP address 155.76.159.253 was from an Army Corps computer. Using logic that most people learn in the sixth grade, these screenshots proved that "stevonawlins"—and his vicious comments—came from the Army Corps. I also told Dennis that there were many more examples available.

After sending the proof about stevonawlins to Dennis, I checked the StatCounter and laughed out loud. Within minutes, I could see from the telltale Army Corps IP addresses that sixteen different people with the Army Corps had logged onto the Levees.org website. And, in doing so, all had shown themselves to be Army Corps personnel closely involved in this scandal. Upon Bruce's urging, I had caught the Army Corps red-handed. Appropriately, the IP address description on the StatCounter—amvn91h.mvn.usace.army.mil—was in bright-red lettering.

The folks at the Army Corps' New Orleans District Public Affairs Office—and those above them—had likely thought they were so clever that they could surreptitiously attack me and remain anonymous. But I could see them. And now their activity would be exposed.

* * *

Later that morning (December 15), I drove to the WWL-TV Channel 4 station, which was located in a terribly inconvenient place: the heart of the Vieux Carré. After a staffer miked me up, Dennis asked questions about what I had discovered. I spoke about how the comments were intended to obscure the facts about the 2005 flood, and how, taken together, they shifted responsibility for the flooding away from the Army Corps and onto local residents.

I also tried to drive home another point; there were hundreds of comments that I had witnessed over the course of several years. Despite not having a way to describe the activity at that time, I knew that these comments were far more than the work of one person called "stevonawlins."

That afternoon, "teasers" appeared on CBS television's local station: "Tonight on *Nightwatch*, Levees.org says it has discovered attacks posted on its blog are coming from inside the Army Corps of Engineers." The news station featured the commenting scandal on both the six and ten o'clock news.[347]

My inbox was instantly flooded with email from supporters, family, and friends, congratulating me on this chilling discovery and expressing disbelief over the Army Corps' Public Affairs officer's response: "There are over 1,600 people working at the Army Corps of Engineers New Orleans District to reduce risk for the Metro New Orleans area. This isolated incident involved one person out of 1,600 people."[348] No apology; just a lofty statement about reducing risk.

We moved fast to further the reach of the WWL-TV Channel 4 news story. Early the next morning (December 16), Francis came to my home and filmed some footage of me talking about the scandal. Then he used his editing magic to create a video that included the WWL-TV Channel 4 story.

A few minutes before four o'clock, I uploaded the video to YouTube and asked our supporters to view it, rate it, and comment on it within twenty-four hours.[349] The campaign had begun! And, once again, Ken McCarthy's tried-and-true formula resulted in our video reaching Top Ten and Top Rated in News & Politics on YouTube, which gave even further reach to the news story.

* * *

Early the next morning (Wednesday, December 16, at 8:06 a.m.), Army Corp district commander Colonel Al Lee sent an email to all senior leaders, asking them to remind their subordinates that any use of government computers to harass a civilian is subject to disciplinary action. Colonel Lee also asked the district's regional information officer in Information Technology if it would be possible to track down who posted the comments to Levees.org's website.[350]

* * *

Thursday, December 18, was a busy day. In addition to monitoring our recent comment-scandal campaign, I had to play a league tennis match at Audubon Park. As busy as I was, I always managed to carve out time for that blessed bit of exercise, which kept me sane and turned big challenges into little ones. I also had a haircut scheduled for later that day.

Rushing between the tennis match and the hair appointment at Magazine and Orange Streets, I spotted an email with this subject heading: "A letter from Col Alvin Lee, commander, Army Corps, New Orleans District." It was sent from Rene Poche with a usace. army.mil email address. I opened the attachment to find a letter on Army Corps letterhead, which began with this sentence:

"Please accept my apology for the unprofessional comments someone in my District posted to your web site."[351]

While it admitted to the misdeeds of only one person, Colonel Lee's apology was a complete about face from the stuffy response the New Orleans District Public Affairs Office had issued three days earlier.

I forwarded the letter to Dennis and dashed off to my hair appointment. Forty-five minutes later, my cell phone rang as I was driving back home. It was a producer for a local radio show, who in a rather rough tone told me that he had seen Colonel Lee's letter of apology on WWL-TV Channel 4's website.

"Oh, I didn't know that they had posted it," I said.

"Can I ask you a couple of questions?" he said.

I replied, "Of course."

In hindsight, he probably audio-recorded my responses, but it didn't matter. By now, I had learned to treat every word I uttered and wrote as public property. That way, I never had anything to worry about.

"Are you going to try and find out the identity of the person who left the comments?" he asked.

I literally had never even thought of that.

"No," I replied. "We want to expose the fact that this action is far more than the work of one person."

"How many hits does the YouTube video have so far?"

I told him that it numbered more than 20,000 the last time I checked.

"You got the video up fast," he said.

"Yes, because it's imperative."

He said thanks and hung up without saying goodbye.

One week later (December 22), our video grossed 25,000 views, and *Gambit* threw a "brick" at the Army Corps for saying that only one person within its ranks had posted vicious comments about Levees.org.[352] The day before Christmas, Colonel Lee requested his information-technology officer to block Levees.org's website from all employees except the Public Affairs Office.[353]

It was a fabulous campaign, and the Army Corps was red-faced. But I was not satisfied. I remained convinced that this was only the tip of the iceberg, but I did not yet know the words to describe what the Army Corps was doing. And, even if I did know the words, I could not prove it. But there was someone who could: Jon Donley.

Chapter

8

Facing Off with the Army Corps

John Donley, editor-in-chief of NOLA.com, had a front-row seat to the sheer scale of anonymous comments radiating from the Army Corps' New Orleans District headquarters. Donley's back-end tools were far more sophisticated than mine. For this reason, I scheduled a meeting with him on Friday, January 23, 2009, at his Poydras Tower office in the central business district.

At the meeting, I got straight to the point. I believed that Levees.org was the target of an organized campaign designed to discredit Levees.org and displace blame for the flooding away from the Army Corps and onto the people of New Orleans. All I needed from Donley was a printout of all the comments coming from IP address 155.75.159.253. I made a point of saying that I did not want to know the senders' identities. I just wanted to know how many different senders there were, how many comments there were, and what they said. My goal was to prove the breadth of the campaign. Donley appeared to understand.

At three thirty, while I watched, Donley picked up the phone and requested that the software technical supervisor produce a printout of all comments emanating from the Army Corps IP address 155.75.159.253. At six thirty that evening, Donley called me to say that he had the printout in his hands, but only for comments from the past six weeks. He and Dwayne Fatheree, his second in command, were looking over it together. According to Donley, in the roughly six-week period alone, there were almost seven hundred comments from IP address 155.75.159.253. Some of them civil, but most were the vicious sort that I had seen.

But there was a hitch. Donley needed to find out if giving me this information was allowed by the newspaper, and we agreed to speak again the following week. Before he hung up, Donley remarked that Colonel Lee's Christmas Eve admonition to his staff to discontinue the bad behavior was clearly being ignored.

"These comments are still coming in at exactly the same pace in January as they were in December," he said.

Over that glorious weekend, I had fantasies of sitting on the floor and poring over reams of paper from an old-fashioned printer where all the pages were connected. I waited with great expectation to hear back from Donley on Monday. But days went by. Then a week. I still did not hear from Donley. He also did not respond to either my emails or phone calls, which was odd.

At the end of the following week (February 5), I was downright worried. This was unlike Donley. On February 10, I wrote again to let him know that I had found a new comment posted to my NOLA.com column from the same Army Corps IP address:

> So what were you people doing living
> BELOW sea level... That's like letting your
> child play in the street and then complaining
> when a speeding car kills him.[354]

* * *

While waiting to hear back from Donley, I reached out to Jed Horne to draw his attention to some errors in Chapter Seventeen of his book *Breach of Faith* which pointed fingers at local levee officials.[355] I took great pains to summarize my concerns in a nine hundred-word email sent on February 27. Horne responded early the next morning with a very brief note starting with "My account threw the Army Corps no roses" and closing with "keep up the good work."[356] He did not address my concerns, but I now had the right to publicize them. I wrote a book review and published my concerns online. I heaped praise on Horne's book but indicated that I had a problem with Chapter Seventeen. I then proceeded to explain that Horne wrongly blamed local officials for the 2005 flood. I closed the review with this: "Fine job! to Horne for his book which is a joy to read. And precisely because this book will likely be read for decades to come, we hope Horne considers updating Chapter Seventeen in future editions."[357]

* * *

On February 14, I got word: Donley and Fatheree had been "let go." A mutual colleague at NOLA.com introduced me via email to Gary Scheets, who would replace Donley.[358] I was devastated. My dreams of poring over thousands of comments would not come true. I felt there was a connection between Donley's firing and his intended cooperation with my request.

I emailed Scheets. My plan was to work with him with the outward appearance that I expected him to let me see the comments coming from the Army Corps headquarters. Scheets suggested that we talk after Mardi Gras. I dutifully waited until after Mardi Gras (which, that year, was on February 24) and emailed him on February 25, March 16, and March 19. No response.

I decided to go over Scheets's head and appeal to his boss, James O'Byrne, for the comments data. After a week, O'Byrne acknowledged my request and informed me that their New York attorney was reviewing its legality.

Three weeks later—on April 15—I received this curt reply from O'Byrne:

> NOLA.com does not release any user information to third parties without a valid subpoena that compels us to do so. I do not possess, nor does anyone on my staff have any knowledge of, any report that may have been promised or prepared regarding the source of comments or any other proprietary user information connected to the levees.org column, which is hosted by NOLA.com as a courtesy to your organization.[359]

After standing on the tip of a mountainous scandal for three months, I had received an answer to a question I did not ask. I requested the comments associated with an Army Corps IP address—data that was already publicly available though not batched together to show its significance. I did not request "user information" which is the commenters' personal information. I wanted to expose the enormity of the activity, not expose anyone personally. This was a deliberate tactic to steer attention away from the critical issue—the Army Corps was rewriting history. But feeling that it was useless to continue, especially now that Jon Donley and Dwayne Fatheree were let go, I chose to focus on other goals of Levees.org.

* * *

On May 5, just as I was preparing for a videotaped interview to be recorded downtown, I received an email from Bruce Biles.

He said only one thing: "Check out my blog."

I did just that. Somehow, Bruce had managed to access a nonpublic webpage from OPP's (Outreach Process Partners) website. According to the webpage, the public relations company bragged that it had lowered the number of negative media stories about the Army Corps' New Orleans District Office. The graphs listed the names of the media outlets. The first name on the list was the *Times-Picayune.*

It is worth reminding the reader that the *Times-Picayune* was the media outlet on the ground covering the 2005 flood right from day one. The *Times-Picayune* is likely the local media outlet on which every national media outlet depended most.

The website boasted:

> OPP fosters strategic relationships with media outlets that result in more accurate and balanced stories.[360]

The webpage included a graph, showing the number, in thousands, of negative stories before OPP and after OPP.[361] In addition to the *Times-Picayune*, the website also listed WWL-TV Channel 4, the Associated Press, *USA Today*, The Weather Channel, *The Today Show*, and *Engineering News-Record.*

Visiting the government watchdog website Fedspending.org, it took about one minute to type in "Outreach Process Partners" and see that the public relations company received one million dollars in 2007[362] and another half million in 2008.[363]

It was now four o'clock, and I was supposed to be downtown in forty-five minutes for a videotaped interview. I was also faced with another even bigger challenge. All the media outlets to which I would normally send a press release about this major find were listed as having a "strategic relationship" with the Army Corps—thanks to OPP.

I decided to turn myself into a media outlet. I quickly took screenshots of OPP's live website and sent an email to our supporters, at that time numbering 23,300. The point I made in my email was this: if the federal government was spending a million dollars a year protecting the Army Corps' reputation, then that same government ought to be comfortable spending a one-time five million dollars for a truly independent investigation to study why the levees failed.[364]

At four thirty that afternoon, the president of OPP emailed a copy of my e-blast to the Army Corps New Orleans District public affairs officers in New Orleans. Ms. Roper-Graham, OPP's president, provided these explanatory talking points:

> The Corps is engaged in billions in new project work. There are many questions regarding this work from a variety of stakeholders as well as regulatory-mandated public meetings. Communications is a standard component of all well-run projects. $1m represents a tiny fraction of the overall budget for these [levee] projects.[365]

Of course, Ms. Roper-Graham avoided addressing the concerns I laid out in my e-blast; a classic strategic move. This was the same technique that James O'Byrne and Lauren Solis and countless other Army Corps sympathizers used: change the subject.

The next day (May 6), Pia Malbran, a producer for CBS national news, emailed me. She had seen a copy of my email to Levees.org's supporters and wanted to do a story. At the conclusion of our interview, Ms. Malbran explained why she was covering the story: "You have asked reasonable questions and made valid points, all backed up with data. That is why you got the attention of CBS national news."

Ms. Malbran's story, which appeared two days later, was excellent. In addition to what I had supplied to her, she had obtained a copy of the contract that the Army Corps had with OPP, which was a three-year contract for whopping $4.7 million.[366] Cain Burdeau with the Associated Press also did an excellent story as did freelancer Georgianne Nienaber.[367] All were national stories.

Meanwhile, every media outlet in New Orleans was quiet as a mouse.

* * *

Buoyed by the national stories about the Army Corps and OPP, I reached out to Jon Donley, who by now had moved to San Antonio, Texas. I convinced him to sign an affidavit describing what he had seen while he was still editor-in-chief of NOLA.com. By the first week of June 2009, I received an affidavit from Donley, describing how about twenty different individuals at the Army Corps' New Orleans headquarters posted hundreds of abusive comments on NOLA.com's user-engagement features.

Donley's final remarks are perhaps the most interesting:

> Due to the sheer mass of postings, I was unable to parse the body of comments with the Corps IP address via the site's internal search tools…such a search repeatedly crashed the admin tool. I did, however, created [sic] a spreadsheet of comment IP addresses on the Breaking News blog that covered a roughly six-week period of late

> 2008 and early 2009. What I found in that period was a body of nearly 700 comments emanating from the Corps IP address, an average rate of about fifteen posts a day… This indicated to me that for more than two years, one or more persons, using multiple aliases and posing as regular New Orleans residents, had been systematically using tax-funded resources in personnel time and federal computer infrastructure to conduct a systematic attack on critics of the Corps of Engineers via NOLA.com's user-engagement features.[368]

On June 18, I sent an email to Levees.org's ever-growing base of supporters, requesting each of them to write to the *Times-Picayune*'s publisher, Jim Amoss, and ask him to release the comments data.[369] On July 28, Amoss sent a single email response in a group email to the fifty or so Levees.org supporters who had written to him.[370] He did not deny that thousands of comments attacked me and my supporters. He just took offense at Levees.org's request for the comments data and provided no reason for refusing to release it.

Amoss offered the same excuses that the Army Corps used to protect itself, specifically he wrote:

> I am a native New Orleanian. I deeply love my city. My parents and my brother lost their houses in Katrina. It took my wife and me two years to restore ours, in the Bayou St. John area. The editor of NOLA.com, James O'Byrne, had to have his Lakeview house demolished after it sat, inundated, for weeks. The notion that either James or I or anyone

> in our news operations would deliberately withhold information lest it upset the Corps is both preposterous and deeply offensive. Therefore, please forgive me if I choose not to respond to or engage in dialogue with the person who is spreading those allegations.[371]

The stated facts about the employees' living in New Orleans and the display of being offended were tricks of the Army Corps. As far as I could tell, the upper leadership at the *Times-Picayune* and the Army Corps had become one.

* * *

Having pushed Amoss as far as he was prepared to go, I made a copy of Donley's sworn affidavit and wrote a letter to US Senator Mary Landrieu, requesting that she order an investigation by the Justice Department based on what I had discovered and what was corroborated by Donley. In the letter, which was signed by all Levees.org board members, I included the fact that Donley was comfortable saying that the comments he observed were in the "thousands over the entire nearly three-year period" since the 2005 flood.[372] We stated that we believed that these comments reflected a coordinated effort to abuse a public forum with the goal of shifting responsibility for the Army Corps' defective floodwalls away from itself. I hand delivered the letter and affidavit on June 23, a very hot summer day.[373]

On that same day, Senator Landrieu's press secretary sent a statement to WWL-TV Channel 4, confirming the senior senator's intent to demand an inquiry. But instead, she was directing her request to the Department of Defense: "I am very concerned by the reports that a number of Corps employees have engaged in a disingenuous campaign to undercut their critics in Louisiana. My staff and I will continue to review these allegations and will urge Pentagon officials to thoroughly review this matter."[374]

I was worried about Landrieu going to the Pentagon since that was "up the same chain" as the Army Corps. But I had to accept that I had done all I could. It was now out of my hands.

That night, starting at 6:39 p.m., I chatted online with Donley about the day's events. Donley agreed that a Pentagon inquiry did not seem to be the best approach. He also volunteered something else. He explained that he intentionally said little about me and Levees.org in his affidavit and in his interview with WWL-TV Channel 4 for a specific reason: "The upper-level staff of the *Times-Picayune* tends to just blow you off. If I had mentioned your name, they would have said, 'Oh, you're listening to her?'"

Donley continued. "Well, evidently, the top brass got with tech support today and with O'Byrne to test my claim that the IP info was open to any columnist like yourself. They proved that I was right. Oh, and the brass verified for O'Byrne that there were indeed seven hundred or so comments in the period I described, and they verified that the posts were coming from inside the Army Corps. This includes the visibility of IPs, tracking the Corps IP address, volume, etc. They're also trying to change the software, so columnists can't see it."

I responded that I was surprised it took that long. And I was surprised that the *Times-Picayune* had not yet canceled my NOLA.com column.[375]

"It wouldn't look good if the *Times-Picayune* shut down your column," Donley replied. Then he added one more thing before signing off: "Man, the only thing any news organization ever has going for it is its credibility."

* * *

Senator Landrieu kept her word.[376] On August 6, 2009, she requested a Department of Defense inquiry into the Army Corps comment scandal.[377] For six weeks, I waited on pins and needles. The fourth anniversary of the 2005 flood came and went. I dreamed of the Pentagon requesting to see the mountain of comments data after which they would publicly denounce the heinous practice of people in a position of public trust hiding their identity, pretending to be impartial observers, and using NOLA.com to attack people.

During this time, I checked in regularly with Donley and asked if the Pentagon had been in touch. He replied negatively each time. Then, far quicker than I expected, the Pentagon completed its investigation. The Department of Defense's inspector general John Crane had closed the inquiry by stating that Army Corps' New Orleans District officials had taken "appropriate actions once informed of the allegations at issue." The report also stated that further review by that agency was not warranted.[378]

The Pentagon had not contacted the *Times-Picayune* IT department and requested the comments data. The Pentagon had not contacted Donley for further questioning. Instead, it appeared that the Pentagon had limited its inquiry to only the comments that I had discovered on my own. This was a classic whitewash: limiting the scope of the inquiry in order to guarantee that no further bad behavior would be found. I was devastated. This was the last chance of ever being able to see the true volume of the comments data and to expose the reprehensible behavior I had witnessed. And, inexplicably, Senator Landrieu was satisfied.

After receiving the Pentagon's report, I was depressed for days. Yet, as I lay my head on my pillow each night, I reminded myself that I had done everything I could. I stayed the course and focused on what needed to be done next. The Army Corps had stumbled and bumbled so many times before, and they were bound to do so again… and I would be watching.

I found closure in Donley's Christmas-season remarks on my *Huffington Post* blog:

> As editor-in-chief of NOLA.com, the online news and community platform of New Orleans, I tracked the way Corps agents subverted our life-saving forums and blogs into a tool to twist the facts about what happened [when the levees broke]… The Corps' activities on NOLA.com sparked an investigation called by Sen. Mary Landrieu. This investigation—conducted by what was essentially an internal review—pinned the blame on an isolated individual. My data, however, was evidence that was not considered; in fact, I was never contacted by any investigator, although I was the key source of these accusations. The investigation was whitewashed, and one can only speculate what motivated Senator Landrieu and the New Orleans news media to accept the results at face value.[379]

I was in full agreement with Donley. I had watched the media, both local and national, give the Army Corps a free pass over and over again despite the growing abundance of data. From day one, the *Times-Picayune* treated me like I was the harasser rather than the harassed. In a *Times-Picayune* September 29, 2009, story about the Pentagon inquiry, their reporter wrote that the outlet "does not release the identities of online commenters" even though I had never asked for that.[380] I believe the misreporting was intentional, with the goal of making me appear to want to intrude into peoples' private lives. The extent of the abusive comments coming from the Army Corps' New Orleans District Office could have been exposed for all to see with a few keystrokes and the click of a mouse, costing the taxpayer nothing. There was no doubt that the Department of Defense did not want the information to become public. And, for some reason, the top brass at the *Times-Picayune* cooperated and Senator Landrieu did not object.

A week later (October 10), while I was shopping at Lakeside Shopping Center in Metairie looking for a birthday present for my sister Melissa, my cell phone rang. It was Lauren Delery, with Adams and Reese, and she had good news. A friend who owned a software tech company said that he had the tools to access the comment data at NOLA.com that came from IP address 155.75.159.253. The sheer volume and content of the comments would tell the story, and the evidence would speak for itself. This was all I needed! I stood outside the Crabtree & Evelyn store and burst into tears of joy!

That night, I was on top of the world, envisioning the onslaught of media rushing to tell the story. But, a day later, I learned that Ms. Delery's friend was mistaken. His team could not access and retrieve the data that he had promised. Naturally, I was devastated again.

* * *

I interviewed Jon Donley in February of 2018. I explained to him that my biggest regret in leading Levees.org was never gaining access to the comments data coming from the Army Corps' New Orleans District office. I asked Donley if he possessed a hard copy of the comments coming from IP 155.76.159.253 from late 2008 and early 2009. He replied that at one time he had a printout that numbered in the thousands. But he did not have a copy in 2018 as it was not his property. He added that, after he was fired from the *Times-Picayune*, an upper-level official of the parent media company, Advance Local, contacted Donley and strongly accused him of spiriting a hard-copy printout of the data from the office.

Donley felt that it was indeed obvious that the *Times-Picayune* was supporting the Army Corps. Donley had been discouraged from following any line that smeared them, including the comment scandal. Donley added that it was particularly painful for him because he saw it as his job to prevent exactly that sort of abuse.

The most chilling piece of information that Donley shared were statements that Peter Kovacs, the managing news editor for the *Times-Picayune*, had made to Donley pertaining to my LTE in December 2008 just as the Army Corps' comment scandal was unfolding.

Donley had featured my LTE rebuttal[381] to another LTE[382] written by an UNO professor—who was sympathetic to the Army Corps—on NOLA.com.

After Donley featured my LTE rebuttal (January 9, 2009), Kovacs called Donley and said, "You're hurting the Army Corps' inner child." Kovacs told Donley that the *Times-Picayune* didn't want to ruffle the Army Corps' feathers and be seen as a whipper. "The Army Corps has been criticized enough already and so the *Times-Picayune* will cut the Army Corps some slack," said Kovacs.

This revelation from Donley reinforced my certainty that the *Times-Picayune* had in its hands evidence that the Army Corps was using comments on forums, articles, and blogs to paint themselves in a positive light, to use comments to spin-doctor information, and to discredit those who disagreed: the hallmark of an Astroturf campaign.

* * *

In November 2009, US District Court Judge Duval ruled that the Army Corps was financially liable for damages caused by its negligent maintenance of the MR-GO (Mississippi River–Gulf Outlet), a little-used navigation channel that worsened the 2005 hurricane's storm surge.[383]

While this would not help the hundreds of thousands of people harmed by the failure of the Army Corps' faulty floodwalls on the 17th Street and London Avenue Canals, it did provide relief to others in the city. For the residents of the Lower Ninth Ward and St. Bernard Parish, this ruling could bring financial remuneration.

The Army Corps appealed, but the US 5th Circuit Court of Appeals upheld Duval's decision.[384] Bottles of champagne were uncorked. If upheld by the US Supreme Court, the ruling could have resulted in residents, businesses, and local governments being reimbursed for damages up to twenty billion dollars.[385] But that would not happen.

In a stunning move, the US 5th Circuit Court of Appeals reversed its decision six months later with little explanation.[386] The reasoning for the dismissal was the same as for the 17th Street Canal suit; while the Army Corps was responsible for the damage, they could not be held financially liable. This decision was devastating to the people of New Orleans for two reasons. The first reason was obvious in that some of the affected people would get no relief. In addition, the lack of financial damages would sustain previously held beliefs by the American people; if there were no damages awarded by the Army Corps, there must have been no wrongdoing.

* * *

On September 18, 2009, a large, thick envelope arrived from FEMA, addressed to me.[387] It was the response to my request under the FOIA six months earlier when I requested the names of the counties that contained levees. I opened the heavy package and scanned the cover letter. Before looking to see where those counties were, I noticed the total figure. The number of people living in counties protected by levees was not 43 percent, as Galloway had told our members of Congress. It was 55 percent. This figure meant that the majority of the American people lived in counties protected by levees. So, with a mission of education, we broadcast the information in an e-blast to our growing database of supporters.

A month later, I got a call from Ezra Boyd,[388] an LSU masters graduate who had worked as a geographer with Dr. van Heerden on his Team Louisiana report. Boyd had an offer for me. On October 16, at a coffee shop on Jefferson Avenue, he explained how he wanted to take my data from FEMA and generate a levee county map, so the citizens at large could see, at a glance in a pictorial, just how common levees were in the United States. He would also do a study on the average wealth and quality of life in those counties. I saw the merits of his idea and handed over our FEMA data to him.

On December 1, Levees.org rolled out Boyd's levee county map. A short time later (December 23), Boyd produced his analysis which concluded that levees more than paid for themselves when their cost was compared to the investment they protected.[389] In my mind, this analysis would quell any doubt that the New Orleans region should be rebuilt, a common criticism still being heard five years after the 2005 flood.

I offered the story and map to Mark Schleifstein with the *Times-Picayune* as an exclusive.[390] I continued to offer exclusives to Schleifstein because 1) he wrote for the media outlet with the largest following, and 2) he appeared to have some clout with the news editors of the *Times-Picayune.* Schleifstein did write an excellent story, which was reprinted in a dozen other journals, including, but limited to, *Insurance Journal Magazine*, *Homeland Security Newswire*, *InfrastructureUSA*, and *Directions Magazine*. Senator Landrieu utilized the graphic in a hearing relating to rising flood-insurance costs.[391] More recently, in March of 2019, a levee board member from Fort

Bend County Texas requested a high-resolution copy of the levee county map because it was a "great presentation of the levee system in the US."

I marvel over the luck I had in uncovering this significant and influential piece of information. It sprung from a cryptic email from Marc Levitan, director of the LSU Hurricane Center, with no details but just an attachment.[392] If I had not read the attachment carefully, and if I had failed to file the request under the FOIA from FEMA, this statistic might have remained in a dusty file cabinet indefinitely. As usual, by asking questions until I received an answer that satisfied me, I got to the facts either on purpose or by accident.

* * *

In December 2009, my phone rang. It was Glenn Corbett, a six feet, four inch professor of fire science at John Jay College of Criminal Justice in New York City. Corbett was an advisor for two women who were doing work similar to mine: Sally Regenhard and Monica Gabrielle. Both women lost close family members in the September 11, 2001, attacks and both refused to accept money that the government offered with the clause that they remain silent. Both chose to demand more information on why the infamous event happened. Like me, these two women refused to believe initial reports and created an organization (Skyscraper Safety Campaign) with a mission of education.

Gentle giant Professor Corbett was to them what H.J. was to me: a technical advisor. And, just like H.J., he had a deep, ethical concern for public safety. He urged me to consider nominating the levee-breach sites to the prestigious National Register of Historic Places (NRHP). Right away, I envisioned federal rangers at the breach sites wearing distinctive, round-topped hats and dark aviator sunglasses, giving tours at the sites of the worst civil-engineering disaster in US history. Professor Corbett warned me that it was a lot of work, but I was already convinced.

In that initial phone conversation, he alerted me that the nominating process began at the state level and that it was critical to engage those key people right from the start. He suggested that I travel to Baton Rouge and meet with Pat Duncan, the chief historian at the Louisiana State Historic Preservation Office (SHPO).

I emailed Pat on June 22, 2010, and met with her one week later at ten thirty in the beautiful office building of the lieutenant governor at 1051 North Third Street on Capitol Hill in Baton Rouge. She was a quiet, dedicated civil servant who loved her job and her dog. On that day (June 29), Pat, as well as her direct superior Nicole Hobson-Morris, assured me that the SHPO did indeed consider the breach sites eligible for the National Register, and they both encouraged Levees.org—verbally that day and a month later in writing—to pursue it.[393]

Thus began the long, tedious process of preparing the formal documents required for the NRHP listing. Pat wanted me to describe the history of the hurricane protection system as well as the history of the entire New Orleans drainage system. I was getting quite an education myself. I also realized that I was in over my head and needed help.

Corbett suggested that I reach out to his friend Mark Barnes, a retired senior archaeologist in the National Register Programs Division in Atlanta, Georgia. In his career, Mark had worked on thousands of nominations. On October 9, H.J. and I held a conference call with Mark, and we decided to hire him for six months at the bargain price of $950. Even though Mark lived in Atlanta, it was easy to work on a single document from two different states. On Wednesday, February 2, 2011, I wrote a press release about retaining

Mark and offered it to Kevin McGill with the Associated Press as an exclusive.[394] McGill accepted the exclusive and did an excellent story.[395]

Mark was super knowledgeable on how the federal government works and trained me well.

"It's important that you maintain a 'neutral speak' whenever you write or speak with any members of the state or federal preservation offices. That is how you will succeed," he advised me.

I spent the year 2011 working on the nomination. During the process, H.J. and I decided to limit the nomination to two breach sites: the 17th Street Canal in the Lakeview neighborhood and the Industrial Canal in the Lower Ninth Ward. These two breaches, though nearly identical in proximate engineering cause, were otherwise quite different:

- Lakeview was as far west as possible; the Lower Ninth Ward was far to the east.
- Lakeview was predominantly white; the Lower Ninth Ward was predominantly black.
- Lakeview contained middle- to upper-income homeowners; the Lower Ninth Ward had lower- to middle-income homeowners.
- Lakeview was within a stone's throw of a region of the city that did not flood at all; the Lower Ninth Ward was surrounded on all sides by neighborhoods that flooded similarly.

According to Dr. Alexandra Lord, chief of the National Landmark Program in 2013, an event—like the levee-breach event—is worthy of listing if it 1) was a lynchpin moment in American history and 2) caused direct and immediate changes to national policy. In preparing the nomination, I discovered that the list of changes to national policy was long:

- Within weeks of the 2005 flood, Congress directed the Army Corps to identify and map all levees that the agency operates, maintains, or inspects—more than 14,000 levee miles.[396] The Army Corps identified 122 levees as at risk of failing in a major flood.[397]
- Communities whose levees received an unacceptable rating were alerted that they needed to fix the problems, including "movement of floodwalls, faulty culverts, animal burrows, erosion, and/or tree growth."[398]
- Congress directed the Army Corps to take a hard look at I-walls nationwide, resulting in the agency identifying more than fifty I-wall levee projects with "potential performance concerns."[399]
- The agency rewrote the guidelines on levee-building for I-walls (like those that failed at the 17th Street and London Avenue Canals).[400]

While there is no evidence that the Orleans Levee District did anything improper before August 2005 in their maintenance, the Army Corps nonetheless mandated more rigorous levee inspections by every local levee board in the entire nation, including:

- Continuous monitoring and quarterly reporting.[401]
- Digitized checklists and global position technology.[402]
- More frequent removal of debris, vegetation, and silt from levees, and more frequent mowing.[403]

Congress also passed the first ever National Levee Safety Act of 2007, which ordered the secretary of the Army to administer reforms and new programs, including a sixteen-member levee-safety committee.[404] There were also alterations to federal policy unrelated to flooding safety. For example, Pets Evacuation and Transportation Standards Acts (PETS)[405] Public Law 109-308, which was signed into law on October 6, 2006, required states seeking FEMA assistance to accommodate pets and service animals in their state plans for evacuating residents facing disasters. The bipartisan bill passed the US House by a margin of 349-29. Had PETS been in place in 2005, Harvey Miller would not have needed to suffer the heartbreak of watching his dog Monet scream in terror when she was placed in a cage and separated from her caregiver. PETS was necessary because, before the flooding, many chose not to evacuate—particularly the elderly—in order to care for their beloved pets and perished with them.

All these policy changes meant that the majority of the nation's people were now safer. Therefore, there can be no doubt that the 2005 levee-breach event was a catalyst for change in American history. Like the Triangle Shirtwaist Factory fire in New York City on March 25, 1911, when the horrific deaths of 146 people prompted improved factory-safety standards,[406] the levee breaches put America onto a different path.

* * *

I traveled to Baton Rouge every three weeks for a year to meet with Pat Duncan and update her on the nomination's progress. As part of the process, Levees.org was required to alert the owners of the 17th Street Canal (the Flood Authority—East) and the Industrial Canal (the Army Corps). And we did so.

In mid-August 2011, H.J. and I were ready for the final step in the nomination process at the state level: to present our nomination to the State Review Committee in Baton Rouge, which was composed of appointed academic personnel. We sent a letter to our supporters who resided in Louisiana—more than six thousand people—and invited them to join us. I told our supporters that we would have our chance to justify listing two levee-breach sites to the NRHP. We planned a press conference for right afterward to announce the panel's vote. We had lined up an impressive and long list of support, including Senator Mary Landrieu, Louisiana Governor Bobby Jindal, and New Orleans Mayor Mitch Landrieu.[407] I closed the email to our supporters with this statement: "Soon we will commemorate an event that the whole world saw on television."[408]

The night before our panel presentation, at 5:56 p.m. on August 17, 2011, I received an email from Pat Duncan. She wanted to postpone our presentation because the Army Corps claimed it had not had an opportunity to review our nomination document—despite its having the document in hand for over a month.[409] H.J. and I, accustomed to such delay tactics from the Army Corps, responded to Pat that the press had already been alerted and that supporters were already planning to make the trip to Baton Rouge the following morning. We asked that the presentation go on as planned. The SHPO agreed.

The next day (August 18, 2011), Ken Holder, the Public Affairs officer for the Army Corps' New Orleans District, drove to Baton Rouge and announced before the presentation began that the Army Corps did not own the Industrial Canal breach site. Levees.org had previously performed a title search showing that the Army Corps did in fact own the site; nonetheless, this was now a technicality.[410] The panel could listen but could not vote. Fortunately, I had hired a professional to videotape the entire presentation, including excellent comments from Bradley Vogel, a fellow with the National Trust for Historic Preservation, and others who had made the long trip to Baton Rouge.[411]

* * *

That same month, on August 10, 2011, the Associated Press issued a style guide for reporters writing about September 11, 2001, and the tenth anniversary of the terrorist attacks in New York City.[412] The reason was a sound one: to assist journalists in developing consistency in their reporting on the disaster that caused 2,752 deaths[413] and changed national policy in America. But the media outlet did not issue a style guide on how to reference the worst civil-engineering disaster in US history, which caused at least 1,577 deaths and triggered significant changes to national policy.[414]

* * *

The second panel presentation before the State Review Committee was set for November 17, 2011. Undeterred, and confident that the process would continue, we had utilized the extra time to gather more letters of support. Three months later in the same room, H.J. and I showed up again and gave another presentation, now quite polished. And we played the videotape of supporter comments. When it was time for the panel to vote, we were surprised that six of the nine governor-appointed board members voted against our nomination. Their reason was that they were "uncomfortable" with our nomination because it faulted the Army Corps for the levee failures at the 17th Street and Industrial Canals.

"You have made this very difficult for us," said chairwoman Glenna Kramer of Franklin. "It's so long and so complicated."[415]

Despite the board's vote of three for, six against, the state's historic preservation officer Pam Breaux and her staff confirmed six weeks later that they would continue to support Levees.org and the eligibility of the breach sites. They would override the vote of the board.[416] Levees.org could not have succeeded to this point without the intellect and support of Pat Duncan, Nicole Hobson-Morris, Pam Breaux, and many others at the Louisiana SHPO.

* * *

On June 15, 2012, Carol Shull, the interim keeper of the NRHP, denied listing the levee-breach sites as historic.[417] In her letter, Ms. Shull asked why we had selected these breaches and not others.[418] This question was odd because, as my hero Dr. Lord had said, to be worthy of nomination, a site does not need to be the best; it just needs to be "one of the best."[419] But most egregious was Ms. Shull's observation that forty-six of our ninety-one citations in the document were "missing" and the pages were not numbered.[420] This was preposterous! It was clear that, when the document was photocopied, the very lowest portion of nearly half the document was unintentionally lopped off. We had spent more than a year working on the pristine document, and we would have been happy to FedEx another copy overnight to Ms. Shull. Or they could have referred to one of the two indestructible DVDs we had submitted, containing the complete document. But Ms. Shull chose to declare our nomination inadequate due to technical error.[421]

After a year of work, H.J. and I were handed a denial due to a copy machine. But we were not discouraged for two reasons. First, the denial was a hurdle, not the end. The sites are official National Register Eligible Properties and shall remain so until we attempt again to have them formally listed. But most importantly, the process of creating the NRHP document had a blindingly bright silver lining. It turned out that not a single moment of work on the nomination was time wasted because the process forced me to pore over and absorb long, technical reports and digest mountains of new data. The research, and the resulting new data, prompted me to ask more questions. And, like before, the more questions I asked, the more material I found and the clearer my focus became.

For example, environmental reporter Bob Marshall, while working for the *Times-Picayune*, had described the infamous annual "drive-by" levee inspections prior to the 2005 flood as a joint effort between the Army Corps, the Orleans Levee District, and the Louisiana Department of Transportation.[422]

But, as the saying goes, there can be only one Indian chief. So, on March 6, 2012, I wrote to Dr. Bea (the Berkeley team cochair) and asked, "Who is in charge of the annual levee inspections?"

Dr. Bea responded, "You might ask Oliver Houck this question…he has the legal-legislative background to respond correctly."[423]

That same day, I wrote to Professor Houck and asked the same question.

He responded, "I do not know that. But the Corps Public Affairs office here should, and I assume will tell you."[424]

Later that day, I discussed this oddity with my colleague Roy Arrigo. I ruminated on the oddity that two star-studded levee experts could not answer what seemed an elementary question. Roy alerted me that, if I visited the Lake Vista Community Center on Spanish Fort Road on the Lakefront, the answer might be there.

Intrigued, I drove the very next day (March 9) to the unusual circular-shaped low building. Once inside, I found what Roy had spoken about. On two walls were forty-five large, heavy plaques, dating from 1959 to 2004. Each plaque contained the identical wording:

> The United States Army Corps of Engineers, New Orleans District: in recognition of the manner in which the maintenance of its levees has been executed and in which its duly constituted officials have assumed responsibility for a part in the Flood Control Program, a rating of Outstanding is hereby tendered for the year, to the Orleans Levee District.[425]

The only thing different in the forty-five plaques was the date and the name of the Army Corps commander. Suddenly, everything was clear! The Army Corps was in charge of the annual inspections of the Orleans Levee District's maintenance! And the Army Corps gave the Orleans Levee District a grade, just like in grammar school. But it was all perverse. Not only did reporters Gordon Russell and Bob Marshall, and the entire levee-board reform campaign have their story wrong, they all had it backward! For seven long years, the people of New Orleans and the rest of the nation had been fed a fairy tale; that the Orleans Levee District checked their levees for problems only once a year for a few hours and then went to lunch, when in truth, it was the Army Corps that checked the Orleans Levee District's maintenance of one hundred miles of levees in just one morning. These certification plaques explained everything.

How had reporters Gordon Russell and Bob Marshall—who both collected Pulitzer prizes for their stories written in 2005—gotten this information so wrong? Perhaps they were confused and had misunderstood what they were pulling from meeting minutes or hearing in interviews. Perhaps one of the reasons that this issue was so misconstrued is the word "district." The Army Corps could be called "the District" and the local level sponsor could be called "the district." Muddying things even further, the word "inspection" was loosely used to describe the local sponsor's daily maintenance as well as the Army Corp's annual audit of that maintenance. The fact that the entire nation wanted to know exactly what went wrong and wanted to know immediately only added to the chaotic environment.

Only two groups understood this process completely: the Army Corps and the Orleans Levee District. The Army Corps had an incentive to be quiet about this erroneous reporting by Russell and Marshall. And the Orleans Levee District did not have millions to spend on public relations consultants (like OPP) to protect its reputation.

Without absolute proof, one can only conjecture that Army Corps' officials high in the hierarchy were working behind the scenes, speaking to the members of Congress. Such whispering has been confirmed by key people. One of them is civil-engineering expert Dr. Rogers.[426] A second is Tim Doody, a certified public accountant who would later serve on the Southeast Louisiana Flood Protection Authority—East.[427] But it would be three more years before these conversations became public knowledge.

I took photos of the forty-five plaques with my iPhone and uploaded them onto Levees.org's website. These photos were hard proof that Congress and the American people had been duped and that the drive-by levee inspections were a red herring.

The Army Corps had not put forth the perverse myth of the drive-by levee inspections. That was the media, looking to break a story six months before the Berkeley team and the Army Corps-sponsored IPET were complete. On the other hand, Army Corps officials said and did nothing while others took the blame for their own "hasty drive-bys." But having this knowledge in my hands meant little. I needed something official, in writing, that explained who was responsible for what regarding levee maintenance and the annual inspections of that maintenance.

An appendix box on page 2-17 of the Woolley Shabman paper mentioned something called "Inspection of Completed Works." After a little more research, I found the documentation: ER 1130-2-530, issued on October 30, 1996, by the Army Corps.[428] This explained that the Army Corps, and only the Army Corps, is in charge of inspecting levee maintenance.

Wasting no time, H.J. and I penned an op-ed about our newly made discovery and sent it to the *Times-Picayune*, which published it.[429] (By this time, the editorial board of the *Times-Picayune,* which publishes op-eds, had developed a cautious respect for Levees.org, but its news bureau continued to treat me and H.J. as personas non grata.)

* * *

There was another silver lining to our NRHP nomination. While doing research, I grew suspicious of the true power of the Flood Authorities—East and West, which were created by the voters in 2006. It was looking to me like no one could tell the Army Corps what to do, including the shiny new Flood Authorities. So, on September 27, 2011, I emailed the president of the Flood Authority—East, Tim Doody.

Copying H.J., I wrote, "We are writing to ask if the Authority might consider clarifying its responsibilities? Does the Authority currently review Army Corps calculations and designs? Does the Authority currently have any role in construction?"[430]

Doody responded pronto: "Sandy—that will require some board discussion. I see the potential for confusion (potential that the rest of the board may not agree with me on however). Our mission was developed as a result of some long meetings and 'wordsmithing.' I'll get back to you."[431]

The Flood Authority—East had been operating for six years, yet the president could not tell me its role without a board meeting. On December 29, 2011, I reached out to another Flood Authority—East board member, John Barry with the same question.

He responded, "Try Steve Mathies, the former senior staff guy at the state Coastal Protection and Restoration Agency (CPRA) who might speak more freely since he's no longer the executive director. And if you find out the answers, let me know, lol."[432]

On January 6, 2012, I met with Steve Mathies, the first ever executive director of the CPRA. Mathies was a large, tall man with a calm, quiet voice. Sitting comfortably at my kitchen table for more than thirty minutes, Mathies confirmed what I already understood; that there is only one local sponsor: the CPRA. I asked him if the local sponsor had the power to stop the Army Corps from building a project. He replied that all the sponsor can do is comment, nothing more.

My suspicions about the power of the new Flood Authorities was confirmed on June 19, 2014, when I attended a Flood Authority—East board meeting in St. Bernard Parish Council Chambers. While President Stephen Estopinal was giving his opening remarks, I snapped to attention because I thought I had just heard Estopinal say, "We're the authority without any authority."

I concentrated my full attention on Estopinal and strained to hear him in the large, noisy chamber.

Estopinal continued, "We were not the local sponsor of the new construction that's been going on with the Corps of Engineers. We were not afforded the opportunity to review, approve, or disapprove plans for the protection system. We were not the partner. And, when we did comment, our observations were usually ignored."

Estopinal then gave a litany of examples where the Army Corps ignored the recommendations of the Flood Authorities. I was almost giddy. Here was the president of the Flood Authority—East speaking candidly about how powerless the Flood Authority—East was. And I could not have handpicked a more credible and respected person to say it. Estopinal was the longest serving Authority commissioner, having served for eight years, was highly respected for his expertise, and was regarded as a warm-hearted man.

H.J. and I wrote an op-ed article with this new information and submitted it to Terri Troncale with the *Times-Picayune*. In our minds, this was a blockbuster. With all the energy, money, and time spent on levee-board reform by the business community, according to Estopinal, the Flood Authorities were bystanders who could only wring their hands.

Troncale told me they could print a shorter version of our piece in the form of an LTE. It seemed that the media was straining to keep the fairy tale alive, namely that the pre-flood Orleans Levee Board was corrupt and the new Flood Authorities were a panacea. I thanked her, and then posted the piece to my NOLA.com column and my *Huffington Post* blog. I reminded myself again that the failure of the press to print our message was not a reflection on our rightness.

A fascinating endnote to this important issue is a quote by John Barry, Flood Authority—East board member, in an October 2014 story in the *New York Times Magazine*. The author wrote, "It soon became clear to Barry that despite its grand mission, the board was, as its president, Stephen Estopinal, liked to joke, an 'authority without any authority.' "[433]

> The Flood Authority—East could not write policy, enforce law, or mandate levee construction. It could serve only as a consultant to other agencies, particularly the Army Corps of Engineers—a body not particularly known for welcoming outside help. The Army Corps invited the levee board to meetings, but the relationship soured when the board's experts raised objections to the Army Corps' flood-protection plans. "They got tired of being criticized," Barry said.[434]

* * *

Being forced to read and do research for the National Register nomination also trained me to look for other cues and clues. For example, in reading the Woolley Shabman paper, I found some key information buried deep in the 333-page report that was absent from the executive summary.

The first significant item I found was that the Army Corps had decided on its own to raise the walls of the 17th Street Canal using steel pilings. The Army Corps was not forced to implement that proposal as it had previously whined. This was key. The report stated that, for reasons unique to that canal, the Army Corps recommended raising the walls for the 17th Street Canal, but then recommended gates for the Orleans and London Avenue Canals because the gates plan was significantly cheaper. Again, this defied common lore, namely that the Orleans Levee Board had pressured the Army Corps for the cheapest option. The fact was, the Orleans Levee Board wanted the more expensive option for the Orleans and London Avenue Canals, the same option as the 17th Street Canal. (They believed it was the better design.)

Just like Gordon Russell's "drive-by levee inspections," the story of the Orleans Levee Board pressuring the Army Corps to build I-walls "because they were cheaper" was not only wrong, but exactly backward. As I always did when I found something important, I picked up the phone and called H.J. But I would soon discover something even bigger in the Woolley Shabman paper.

In the 1980s, having just been chided by the GAO, the Army Corps was behind schedule and faced with rising costs. In response, General Tom Sands, from the Army Corps' Mississippi Valley Division headquarters, ordered a large-scale "Sheet Pile Wall Field Load Test" (the E-99 Study) in the Atchafalaya Basin, a region with soft clayey soils similar to New Orleans. Tragically, the Army Corps engineers misinterpreted the results of their study and wrongly concluded that, when foundation soils were poor, sheet-pile penetration depth beyond a certain point would not significantly increase I-wall stability.[435] In other words, the Army Corps determined that it only needed to drive sheet piles down to depths of not more than sixteen feet instead of between thirty-one and forty-six feet.[436] Beyond that, it was considered a wasteful expenditure.

In December 1987, the Division Headquarters issued new criteria guidance to the New Orleans District on sheet-piling design based on the E-99 Study tests.[437] The phase-in of the new I-wall criteria was embraced because of the "high potential for savings."[438] Records show that the switch to the shorter sheet piles saved approximately $100 million.[439] However, these cost savings would come, eighteen years later, at the expense of engineering reliability.[440]

This was excruciatingly important. On August 29, 2005, the breach of the 17th Street Canal, combined with the two breaches of the London Avenue Canal, caused at least twenty-seven billion dollars combined in direct residential, commercial, and public-property damage in the city's main basin.[441] Put simply, the failure of the outfall canals was due directly to an egregious engineering error that the Army Corps had made in the 1980s—on its newest floodwalls—in order to save money on steel.

This detail was not included in the Woolley Shabman paper's executive summary! This omission, along with the omission that the Army Corps elected on its own to raise the walls of the 17th Street Canal, were both news to me—someone who was extremely familiar with the disaster and its causes. Although the Woolley Shabman paper itself was superb, the executive summary appeared edited by a third person who perhaps did not want the damaging details about the Army Corps being picked up by the press. The Woolley Shabman paper was released in July 2007, and, like most everyone else at that time, I read the only executive summary.[442]

Without reading the entire report in May 2011, I would not have discovered these details. The Army Corps' choice against building a gate at the 17th Street Canal and the wrong conclusion to the E-99 Study were extraordinarily important—and were omitted from the executive summary. John Barry and Dr. Ivor van Heerden, both highly knowledgeable experts, were surprised when I brought these facts to their attention.

Chapter

9

Bayoneting the Wounded

Ever since the 2005 flood, I was the target of a constant stream of abuse, disparagement, and even slander in the *Times-Picayune's* community-engagement features. In addition to the people I had already unmasked using the Army Corps' IP address (users "stevonawlins," "swain," and "overwrought"), there were others ("honestred," "beestung," "truthexposed," and "heidihoe"). Comments that labeled me a liar were allowed to stay visible online with the *Times-Picayune* until as recently as February 2013.[443]

One persistent commenter harassed me from October of 2005 until 2014, using a modification of the initial username "moderator," including "moderate1," "mod1," "moderation1," and eight other derivations using the letter "l" or the number "1." All of them shared a common theme, namely that the Orleans Levee Board deserved blame for the 2005 flood.

By late 2012, having been the butt of vicious remarks at the *Times-Picayune* for seven years, it was not easy to get my attention. But one of the derivatives of "moderator" had begun badgering me in a new fashion. This commenter "moderatel" had managed to find an e-blast that I had issued more than six years earlier to Levees.org's supporters in which I stated that Garret Graves with Senator David Vitter's office invited me to submit testimony to the Senate Homeland Security and Government Affairs Committee.[444]

In a case like this, the accurate legal term was "statement." But Graves had used the word "testimony" in his invitation, and so I used it in my e-blast. Now, "moderatel" was accusing me of lying about the incident six years ago. (My high crime was calling my submission "testimony" when it was really a "statement.")[445] But something much more important than that caught my attention. The commenter had cut and pasted the conclusion from the Berkeley team report into the comment as his supporting data:

> The USACE had tried for many years to obtain authorization to install floodgates at the north ends of the three drainage canals that could be closed to prevent storm surges from raising the water levels within the canals. That would have been the superior technical solution. Dysfunctional interaction between the local Levee Board (who were responsible for levees and floodwalls, etc.) and the local Water and Sewerage Board (who were responsible for pumping water from the city via the drainage canals) prevented the installation of these gates, however, and as a result many miles of the sides of these three canals had instead to be lined with levees and floodwalls.[446]

In a thunderclap, I realized something that I had missed before. The Berkeley team report's conclusion was the only credible expert conclusion that faulted the pre-flood Orleans Levee Board. As long as this conclusion existed, specifically unchallenged by other experts, no amount of new, conflicting evidence was going to matter.

I had already seen that new evidence did not erase the sentimental and wrong story that most Americans carried around in their heads—a monster storm had drowned a corrupted city below sea level. Big media outlets would continue to cling to its original "fairy tale" script. This was certainly true for the *Times-Picayune* whose reporters had collected Pulitzer prizes for their wrong stories.

I called H.J. and explained to him that our big roadblock was the conclusion of the Berkeley team report. As long as that conclusion went unchecked, our work would never be done.

H.J. understood and agreed. But what would we do about it?

The answer to that question was our biggest challenge since the formation of Levees.org.

Chapter

10

A Major Coup

Bit by bit, Levees.org chiseled away at the web of interwoven myths that created the fairytale of the 2005 flood, namely that New Orleans was deluged by a monster storm worsened by local corruption and the city's below-sea-level geography. Levees.org had found these startling facts:

- Two Pulitzer prize-winning articles about the drive-by levee inspections were exactly backward. It was the Army Corps that did its annual levee inspections in one morning, not the Orleans Levee District.
- Three crucially important details were absent from the executive summary of the seminal Woolley Shabman paper funded by the federal government about the 17th Street Canal. Specifically, it omitted that 1) the Army Corps selected, on its own, to build the "high wall plan" for the 17th Street Canal, 2) that plan cost the same as the "gates plan" and 3) that plan was not a "fallback plan" as multiple top level Army Corps spokespersons had claimed.[447]

- Most importantly, the Woolley Shabman paper's executive summary omitted that the Army Corps, in looking for ways to save money on steel, had misinterpreted the results of its own study. This mistake doomed the city because it affected the design of the floodwalls on the 17th Street Canal and also the London Avenue Canal and the Industrial Canal. The gravity of these omissions cannot be overstated.

But Levees.org's most important discovery was the Berkeley team report's wrong conclusion, arrived at before May 22, 2006. The Berkeley team report, funded in part by the NSF, had been held aloft as watershed proof that the local pre-flood Orleans Levee Board should be held partly responsible for flooding. This unsupported conclusion appeared in an early major investigation, completed on a shoestring budget, and without the benefit of time to do a complete study. Furthermore, the unsupported conclusion fit neatly into a pre-established popular belief: that the people of New Orleans were corrupt and therefore deserved their fate.

These disconnected thoughts were swirling through my head as I was posting an online comment to a February 2013 *Times-Picayune* story.[448] Garret Graves was testifying before Congress that the Army Corps retains exclusive control over building Greater New Orleans's hurricane flood protection, "in addition to general immunity, while the nonfederal sponsor's role is largely reduced to that of a bystander."[449]

In my comment, I wrote, "Lt Gen Strock admitted to a reporter for the *Times-Picayune* that when he blamed local levee boards for the New Orleans flood, he relied on things he had heard but not personally researched. It's been seven years and no investigation has found evidence that the levee boards are responsible for the flooding."[450]

The now-familiar commenter "moderatel" promptly replied: "That's a lie and you know it. And no matter how many times you repeat it does not make it true. The truth is the initial report from the National Science Foundation [the Berkeley team] says it clearly, and you know it. From Chapter 15 (in quotes): The locals forced those canals of the flood system that created most of the failure."[451]

If I ever find out who this commenter is, I will thank him/her because the comment brought something to my attention that I had missed.

* * *

If one were to look at the history of the levee-board reform, one would be challenged to find even one civil-engineering expert who advocated for it:

- The initial advocate was the New Orleans business community.[452]
- The second advocate was Mayor Nagin's BNOBC, even though the commission's expert panel on flooding did not advocate for the reform.[453]
- The third advocate was the media, most especially Bob Marshall and Jed Horne,[454] who both had collected Pulitzer prizes for their disaster coverage while working for the *Times-Picayune*.[455]

As far as I could ascertain, the only civil-engineering experts who had criticized the pre-flood Orleans Levee Board were the three cochairs of the Berkeley team.

There is an academic axiom that all conclusions are considered preliminary until further investigation. Put another way, all scholarship is considered true until new material comes along. But my experience was that big media outlets, famous authors, and politicians ignore this common-sense rule. Any new material, which laid blame at the feet of the Army Corps and exonerated local officials must come in the form of a mushroom cloud. The new material had to come from a person or a group of people who were extraordinarily credible, and it had to be introduced through a medium that was equally extraordinarily credible.

H.J. and I came up with an idea. We wanted the Berkeley team cochairs to retract their wrong conclusion and replace it with the right conclusion. We got to work straight away on that goal.

* * *

Both H.J. and I had a good dialogue with Dr. Bea, having communicated with him many times since January 2006. With a copy to H.J., I wrote to Dr. Bea and let him know of our concerns about the Berkeley team report. Dr. Bea responded that he was too busy to help us because at that time, he was an expert witness for the last remaining case with the levee-breach lawsuit.

Undaunted, I asked H.J. if he would email Dr. Bea directly since they both were civil engineers and had that magic-brotherhood connection. This time, Dr. Bea agreed to talk to us.[456]

On Thursday morning (March 14, 2013), H.J. and I conducted the phone call from my office with my landline and General Electric speakerphone. The conversation started well enough. Dr. Bea was enthusiastic about speaking to us; it turned out he had a favor to ask. He wanted Levees.org to assist him in finding stakeholder leaders who would support a Supreme Court amicus brief. The goal was to urge the US Supreme Court to review the odd, flip-flopping decision of the US Court of Appeals for the 5th Circuit on the levee-breach lawsuit. We saw no problem assisting Dr. Bea with this request.

Then we explained, as gently as we could, that the Berkeley team report's conclusion contained an error that had the end result of blaming New Orleans officials for the 2005 flood. We wanted Dr. Bea and the other cochairs to submit a letter to the NSF, retracting the wrong conclusion.

Dr. Bea, always the warm, friendly man, said that he would be happy to talk about it. So far, so good. We explained that the offending language was, erroneously, about how a dysfunctional interaction between the Orleans Levee Board and the S&WB (New Orleans Sewerage & Water Board) prevented installation of the superior solution.

"I didn't write that," responded Dr. Bea. "Dave Rogers wrote that."

H.J. and I looked at each other, not knowing quite what to say.

Dr. Bea broke the silence. "What you should do is look through the Berkeley team report and locate the paragraphs that bedevil," he said. "Then get Dave to rewrite the section, and I will sign off on it."

The conversation could not have gone better. It was one of the sunniest days that H.J. and I had since the levees broke. But we were now faced with a new challenge. Neither of us had a relationship with Dr. Rogers, who taught geological engineering at Missouri University of Science and Technology in Rolla. As we mulled over the challenge of reaching Dr. Rogers, I attended a nongovernmental organization (NGO) meeting at the Army Corps headquarters on Leake Avenue on the afternoon of that very same overcast, chilly day.

These NGO meetings were one of the best examples I had seen of the Army Corps attempting to build good relationships with the community that it served. The meetings began at the urging of Public Affairs officer Ken Holder, who had the support of Colonel Ed Fleming. Attendants, like myself, could request that a topic be included on the meeting agenda and could make a presentation to the Army Corps and the rest of the group. At this maiden meeting, I chose to just watch and observe, still feeling gun-shy over past treatment that I—and supporters of Levees.org—had received from the Army Corps.

When the meeting drew to a close, I headed back to my car in the enormous parking lot on the Mississippi River levee. Just then, Paul Kemp (coauthor of the Team Louisiana report) walked by on the way to his vehicle. I had not seen him since the ASCE meeting with Larry Roth. I mentioned my conversation with Dr. Bea that morning, and the challenge of reaching out to a nationally renowned expert in another state whom we did not know.

To my delight, Paul offered to introduce us. He was currently working with Dr. Rogers on a paper. But before the introduction, he wanted to see what H.J. and I had found. I promised that I would send our material to him.

As I turned to walk to my car, I was face to face with a gigantic shipping vessel floating past on the great Mississippi River. I watched in awe for a few more minutes before getting into my car and passing through the security gate. That night, I emailed Paul our treasure trove of discoveries. We set up a meeting the very next day to discuss it.

The sky on the following day (March 15) was bright and blue. I had some difficulty finding PJ's Coffee on Leon C Simon Drive and Franklin Avenue. It was newly opened, and the sign was not yet erected. But I found PJ's and Paul Kemp, and we got straight to work. Kemp was a recently appointed commissioner on the new Flood Authority—East and seemed very willing to help. And he wanted our help as well. Like Dr. Bea, Paul wanted Levees.org to recruit stakeholder leaders to support the amicus brief. I agreed to help. When we parted ways, Paul told me that he would be meeting Dr. Rogers later in the month and would be sure to discuss our project.

* * *

I spent the next two weeks gathering support for the amicus brief to assist Dr. Bea and Paul Kemp. I contacted the Gulf Restoration Network, Glenn Corbett at John Jay College of Criminal Justice, the National Wildlife Federation, and Harry Shearer—all of whom agreed to support it. I contacted the Environmental Defense Fund and the Water Protection Network and received no responses. (On Monday, June 23, 2013, the Supreme Court refused to review the September 2012 decision by the US 5th Circuit Court of Appeals. More than five years later, the final lawsuit was dismissed.[457])

Sure enough, in late March 2013, I arrived home and found a message on my answer machine from Paul, letting me know that he had spoken with Dr. Rogers, who was expecting to hear from us. H.J. and I were excited about the idea of a retraction, but we also started thinking about the idea of a whole new journal article devoted to correcting wrong information that had become entrenched as common household knowledge.

But an enormous challenge still stared us in the face. We needed to convince Dr. Rogers he had made a mistake that he needed to fix. This challenge kept me awake at night. Then, I thought of something that might work. What if H.J. and I made the 670-mile journey to Dr. Rogers's office for a face-to-face meeting? This effort would demonstrate that we were committed, dedicated, and serious about our request.

I talked to H.J. about traveling to Rolla. He mulled it over and came up with an idea. The Rolla campus was just one and a half hours from where his daughter attended college—Washington University in St. Louis. On May 9, 2013, H.J. was due to fetch his daughter and drive her home for the summer. H.J. suggested that I fly up on the morning of May 9, he would fetch me at the airport, we could meet Dr. Rogers, he would drive me back to the airport, and then he would collect his daughter.

So on April 6, I alerted Paul that we wanted to meet Dr. Rogers at his office on May 9. Two days later, Paul responded, "I spoke with him this morning and he is available on May 9, all day with the exception of an 11 a.m. class, and he would be happy to meet with you and H.J."[458]

For the next two days, H.J. and I worked on a letter that outlined our general concerns, and then sent it off. Dr. Rogers responded with this: "I have no recollection whatsoever regarding what the role(s) of the Orleans Levee Board may have been in all of this, other than the supposition I harbor that they increasingly depended on the Army Corps for all engineering issues and assessments, which we would assume THEY had to have previously performed themselves, prior to 1955."[459]

Dr. Rogers was correct. The Orleans Levee Board did indeed rely on the Army Corps, an agency with two centuries of experience which had built tens of thousands of miles of levees. In addition to the email, Dr. Rogers sent a link to the entire unabridged version of the chapter that provided the source details leading to the Berkeley team report's wrong conclusion. He also sent a PDF of an unabridged version of Chapter 4 submitted as a standalone article to the *Journal of Geotechnical and Geoenvironmental Engineering* in May 2008.[460]

I reviewed both papers and found that many of Dr. Rogers's foundational and supporting facts were wrong—so wrong and convoluted that I had difficulty figuring how to respond to them. H.J. and I decided against pointing out errors in the chapter, feeling that that would be neither constructive nor effective. Instead, we would show Dr. Rogers the new data and explain how it conflicted with the Berkeley team report's conclusion.

We had one month to prepare.

* * *

On April 16, Dr. Rogers sent H.J. and me a polite email, explaining how to get to the Rolla campus. He offered to take us to lunch near his office and closed his email with this advice: "It's pretty simple, this is small town, very similar to the fictional one on *The Andy Griffith Show* in the 1960s."[461]

With our face-to-face all set up, I now lay awake at night, fretting over how to approach Dr. Rogers about the errors in the Berkeley team report's conclusion. He had an impeccable national reputation and was the go-to for all kinds of geotechnical catastrophes, including later the Oso, Washington, landslide[462] in March 2014 and the California Oroville Dam Spillway failure[463] in February 2017. The more I thought about Dr. Rogers, the more "bigger than life" he became. I decided to call my mother-in-law Sandra Rosenthal for help. Dr. Rosenthal—or Mamma, as I called her—had, in the course of her career, authored eleven books and nearly two hundred academic articles. Drs. Rosenthal and Rogers were both famous in their fields. I asked Mamma how she thought I should approach him.

It took Mamma just seconds to come up with the answer: "Blame the context. It was an environment of chaos and confusion. It was simply not possible for anyone to fully wrap their heads around the full scope of the disaster so soon."

It was a superb suggestion, especially when you factor in the unreasonable deadline that Dr. Rogers faced on a shoestring budget.

* * *

The day of the meeting with Dr. Rogers arrived. I rose early because my flight departed New Orleans at eight o'clock. I arrived at St. Louis Lambert International Airport at 9:50 a.m., and H.J.—always the punctual one—was there to meet me in his pickup truck. During the drive, we did what we always do: talk about work.

When we arrived at the beautiful, little town, we marveled that it did look like *The Andy Griffith Show*. As we wound our way through the leafy, rural campus we noticed that there were large polyethylene pipes sticking out of the ground in several places. H.J. understood that the university was installing a geothermal heating and cooling system.

We followed the campus map to VH McNutt Hall, arriving just fifteen minutes before Dr. Rogers told us that his lecture would end. We milled around a bit, admiring the immense bronze statue of a miner and his pick. Promptly at 11:50 a.m., Dr. Rogers stepped into the atrium with a few students in tow, asking questions. He soon disengaged himself and approached us with an outstretched hand. Dr. Rogers had a thick head of blondish graying hair and a matching thick mustache. He was of medium height with a solid build and a friendly face and smile. Dr. Rogers asked if we were hungry, which of course we were.

"Do you like sushi?" he asked.

I said yes immediately, and H.J., who does not eat raw fish, said, "Sure!"

As we walked to Dr. Rogers's car, he and H.J. talked on and on about the new geothermal system. I lagged behind to allow Dr. Rogers to get to know H.J. and see that he was a thoroughly knowledgeable civil engineer. Everything was going smoothly, I said to myself as I took in the beauty of the campus in full spring bloom.

At Mottomo restaurant on Kings Highway, three minutes away, we took our seats in the tiny restaurant where, other than an occasional takeout order, we were the only customers. I selected raw shrimp sashimi and a California roll, and offered to split a bowl of edamame with the gentlemen. H.J. ordered chicken teriyaki. Dr. Rogers ordered miso soup and a Rolla Roll—eel and cucumber wrapped in seaweed.[464] I tucked my notebook beneath the table, trying to make the encounter as friendly and pleasant as possible.

After the soft drinks were served, Dr. Rogers leveled his gaze on H.J. and asked how he could help. It was natural to focus on H.J. after all that talk of geothermal magic. H.J., caught slightly off guard, started to talk about the wrong sentences in the Berkeley team report's conclusion that worried us. I gently interrupted him.

H.J. stopped, turned to me, and said—respectfully—"Perhaps Sandy should start."

I turned to Dr. Rogers, waved my hands in the air, and wrung them. Without perjury, I said, "The weeks and months after the levees broke were a chaotic time. It was a context of shock and confusion. There were more than fifty levee breaches. And here was a small group of people—the Berkeley team—on a shoestring budget under an insane deadline with the entire American population demanding answers right then and there. No human could possibly wrap their head around such a complex problem as the levee-breach event in this sort of atmosphere."

As I spoke, Dr. Rogers's eyes flickered in agreement. He nodded slightly.

Encouraged, I continued, "The news media was hounding you, and the Army Corps was reluctant to allow you at the breach sites. It was an environment of chaos through and through."

At this point, I turned to H.J. and urged him to continue. H.J. resumed his conversation and brought up the new evidence that had appeared in an April 2008 Woolley Shabman report, which contradicted a conclusion in the Berkeley team report, specifically relating to the pre-flood Orleans Levee Board.

"Conclusions are preliminary until new investigations introduce new data," offered Dr. Rogers.

We both quickly agreed. Yes indeed! But there was a problem, H.J. said. The conclusion, until it was aggressively addressed head on, would have the end result of "throwing the people of New Orleans under the bus" because it faulted local elected and appointed officials.

To this, Dr. Rogers agreed. "Yes, it does."

I felt that it was time to pull out the notebook. I handed Dr. Rogers a piece of paper with the bedeviling paragraph, which faulted local human beings in New Orleans:

> The USACE had tried for many years to obtain authorization to install floodgates at the north ends of the three drainage canals that could be closed to prevent storm surges from raising the water levels within the canals. That would have been the superior technical solution. Dysfunctional interaction between the local Levee Board (who were responsible for levees and floodwalls, etc.) and the local Water and Sewer Board (who were responsible for pumping water from the city via the drainage canals) prevented the installation of these gates, however, and as a result many miles of the sides of these canals had instead to be lined with levees and floodwalls.

"I didn't write that conclusion," said Dr. Rogers. "Bob Bea wrote that."

H.J. and I, speechless, turned to each other and resisted the temptation to laugh.

"Dr. Bea says you wrote that conclusion," said H.J.

"No," said Dr. Rogers. "That's Bob's style. I can tell."

For seven years, this bedeviling paragraph had the effect of pointing fingers at New Orleans officials, yet neither of the Berkeley team cochairs would own it. The paragraph was the whole reason that H.J. and I had made the seventy-eight-mile trip to Rolla.

Preferring to stay constructive, H.J. and discussed what could be done to address the erroneous conclusion until the sushi arrived. We offered to assist Dr. Rogers with research for a new paper that would right the wrong information about the Orleans Levee Board's perceived role in the floodwall failures of the 17th Street Canal.

"But, if we do that, and succeed in getting this paper published, how will anyone find out about it?" asked Dr. Rogers.

At this, H.J. and I allowed ourselves a full belly laugh. "Don't worry about that," we assured him. "We will take care of that."

Our food arrived. As H.J. fiddled with his chopsticks, Dr. Rogers and I attacked our sushi. During the meal, we discussed other sorts of world issues and discussed our families too. Later, as Dr. Rogers drove us back to campus, he took us on a short tour of Rolla. When he dropped us off at H.J.'s truck, we repeated how important this new paper would be for the people of New Orleans and how pleased we were that Dr. Rogers understood the gravity of the erroneous conclusion and would take steps to address it. Dr. Rogers assured us that he would reach out to the third Berkeley team cochair—Dr. Seed—and get his buy-in too.

With the last detail sealed, we got in the truck and headed back to St. Louis Lambert International Airport in high spirits. We had gotten the green light from Dr. Rogers!

* * *

A day later (May 10), I put together the material I had collected while researching the NRHP nomination. There were events—which took place within three years of each other in the 1980s—that the new paper needed to put it forth as clear as glass.

The first big item was the heretofore unknown fact that the Army Corps had recommended raising the canal floodwalls for the 17th Street Canal but recommended gated structures without pumps for the Orleans and London Avenue Canals because the latter plan was a lot less expensive. Until now, Congress and the American people were led to believe three lies about these canals:

1. The Army Corps wanted to build gates with pumps for all the canals.
2. The gates plan was better and more expensive.
3. The Orleans Levee Board had forced the Army Corps, using political influence, to build a different plan that the federal agency believed was inferior.

Arguably, this was all very technical stuff, which made it even easier for the Army Corps to fool Congress and the American people for years and years. This was exactly why a technical paper, written by technical experts, with a pristinely simple conclusion was necessary.

A second critically important point needed to be made clear. The Army Corps had made a critical mistake in the 1980s. This was the error the Army Corps made during their large-scale test looking for ways to use less steel and thereby save money. These cost savings resulted in floodwalls that would fail when the surge was still far below the tops of the walls.[465] Spoken simply, the walls broke at 50 percent of the load they were designed to hold."[466]

H.J. and I wanted to drive home another point; it is not fair to judge human beings in the 1980s by twenty-first-century knowledge. After the 2005 flood, it was universally agreed upon by engineering experts that exposing the many miles of drainage canal walls in New Orleans to hurricane surge was unwise. Flood Authority—East commissioner John Barry enjoyed saying, if the pre-flood Orleans

Levee Board had expert talent on its board, such as the Flood Authority—East does today, they would have disagreed with the Army Corps' recommendation to raise the walls.

But Barry was judging the actions of human beings at a time when the Army Corps was still revered for doing everything "too good." This was the federal agency that had previously built tens of thousands of miles of levees along the Ohio, Missouri, and Mississippi Rivers, capable of withstanding surge heights of up to twenty feet for thirty days or more every single year. It was not reasonable to expect the levee board members in 1985 to question the competence of the Army Corps. In contrast, it was reasonable for the pre-flood levee board members to consider the Army Corps fully capable of building floodwalls atop a few miles of drainage canals that could withstand about nine feet of storm surge for a period of a few hours and only occasionally, not annually.

The final key point we wanted to include was the danger of making assumptions in haste. After the 2005 flood, the Army Corps used "I told you so" logic to defend itself by saying that it had wanted to build gates for the London Avenue drainage canal but was ordered by Congress to raise the canal walls instead. When, on August 29, 2005, the canal walls fell over, the Army Corps said, in essence, "See? We wanted to build gates, so that means it's not our fault the floodwalls broke!"

The Army Corps touted this defense for years and gained traction with it. But the idea that the Army Corps' gates would have worked properly was an unreasonable assumption. There were fifty breaches in the levee-protection system on August 29, 2005. Any attempt to blame the Orleans Levee Board for the Army Corps' failed design was rooted in the assumption that the Army Corps could have competently built its gates. There were reasons to doubt this. The gates had failed in a much-publicized model test in 1987 at the Army

Corps' Waterways Experiment Station in Vicksburg, Mississippi.[467] It was an embarrassing moment for the Army Corps.

* * *

Bruce Feingerts—the Orleans Levee Board attorney who was heavily quoted in early news stories after the 2005 flood—was a key person to interview before starting a first draft for Dr. Rogers. Feingerts led the campaign to convince Congress to order the Army Corps to increase the heights of floodwalls along the London and Orleans Avenue Canals instead of building gates. Dr. Rogers wrote in an email about Feingerts, "It's either true or he's lying, not like the rest of us who are basically assuming that what we've been told, third-hand, is actually factual."[468]

According to the fairytale, repeated by every news source in the land, Feingerts was the mastermind of a devious plan to create legislation requiring the Army Corps to build cheap, high walls instead of pricey, peripheral gates. This fairy tale was first described in a Christmas Day *Los Angeles Times* article in 2005—a time when the nation was engaged but long before the investigative studies were completed.[469]

I reached out to Feingerts. He agreed to meet me on May 28, 2013, at CC's Coffee House on Esplanade Avenue. I brought H.J. along for this particular meeting. At the appointed time, Feingerts was already waiting for us with a large mug of foamy café au lait. We sat down and explained that we had met with Dr. Rogers, and that we wanted to work as a team to dispel the damaging myths that we believed were hurting New Orleans.

"I was terribly misquoted in the *Los Angeles Times*," Feingerts blurted out.

Of course, we knew which story he was talking about. Feingerts explained to us what he was trying to do. Later, he put it in writing as well:

> I was working the legislative process on behalf of the Orleans Levee Board and brought our engineers and board members to DC to accomplish our goals to best protect our people. I was merely the head of the strategic congressional legislative effort to push for getting the most federal aid participation we could get to properly fund a stronger and more effective flood protective plan to protect against the so-called standard project hurricane event.[470]

For the rest of the meeting, we went over some details that were still unclear to us. From our experience and research, we knew that it was impossible to tell the Army Corps what to do, yet Feingerts had achieved that very thing. We wanted to know how.

We already knew that, by 1990, the Army Corps had decided, on its own, to raise the height of the 17th Street Canal's floodwalls.[471] The Orleans Levee Board and the S&WB loved this plan because 1) they considered it the superior plan, and 2) the feds would pay for it. However, for the far smaller capacity London and Orleans Avenue Canals, the Army Corps stood by its preference for the gates plan (with no pumps) because it was way cheaper.[472]

Feingerts explained how he succeeded in a "power play" with the Army Corps. In 1990, Feingerts suggested to members of the Louisiana delegation that they, in conferencing, insert a ruling or some "language" into an engrossed amendment of the approved version of the 1990 WRDA.[473] The rewritten language redefined the Orleans and London Avenue Canals as part of the Hurricane Protection Project, which were therefore now a federal responsibility. It worked. The 1990 WRDA was signed by President George H. W. Bush and passed into law. The Orleans Levee Board had succeeded in getting the authorization for a project that they believed would keep New Orleans residents safe, and the feds would pay 70 percent.

A week later, I would find written confirmation of this bit of midnight magic in Orleans Levee Board meeting minutes from May 30, 1991. On the bit o' magic, Senator Johnston, in attendance at the meeting, would opine, "The more important things that are done, the things that touch people's lives and pocketbooks, and the things that really measure the effectiveness of what Congress does as far as people in the State of Louisiana, are done quietly, and are done by virtue of knowing what you are doing or the position of power that you are in."[474]

* * *

For one entire month, I delved deeply into the Orleans Levee Board meeting minutes. This was insanely tedious because the search engines to locate key data were useless. These documents had been scanned by Orleans Levee District staffers during the months after the 2005 flood for the congressional hearings. These PDFs had to be read without the benefit of technology. Worse than that, not all the minutes were uploaded online. But it was okay. During that time, dusty file cabinets became my friend because they contained unheard-of details just waiting to be discovered.

For example, I learned that, in the 1980s and 1990s, the Orleans Levee Board had an Engineering Committee, which had oversight of flood protection headed by board member Jerome Dickhaus. The meeting minutes—often exceeding fifty pages—revealed a committee that was highly focused, contrary to State Senator Boasso's and the reform campaign's mantra that the Orleans Levee Board did not pay attention to flood protection.

Perhaps the most important discovery was contained in the Orleans Levee Board meeting minutes for November 7, 1990. These particular minutes were nearly undiscoverable because they were attached as an addendum. On that day, the Army Corps did a presentation on flood protection for the 17th Street and London Avenue Canal. Four people from the Army Corps were present including Colonel Richard V. Gorski and Major H. E. Manuel. The presentation was about the difference between the parallel-protection (high-wall) plan versus the gate plan. The big takeaway was that the Army Corps stated that its preferred plan—the gates plan—would not hinder drainage and that they could not justify the added expense of building higher walls.[475] In other words, the Army Corps wanted the gates plan because it was cheaper, not because it was better.

* * *

Three weeks after our initial meeting with Dr. Rogers, I had prepared a first draft. After H.J.'s review and edit, I sent the draft to Dr. Rogers on May 30. I copied Dr. Bea as he had requested when the project began two months earlier. Four weeks passed, and we did not hear from Dr. Rogers despite several emails to inform him that we had sent the first draft. Another two weeks passed. I got nervous. What if Dr. Rogers had changed his mind? After all, there was no crowd of people banging on Dr. Rogers's door, begging him to retract his statements from seven years ago.

Finally, on July 16, H.J. received a short note from Dr. Rogers. "We have my brother and sister in law coming in this afternoon to visit, for first time in over twelve years. I won't be back in circulation till next week."

Three weeks later (August 7), Dr. Rogers sent us a heavily edited draft. Some of the edits were for readability and flow, but the majority took my layperson's language and changed it into civil engineers' language. Also, Dr. Rogers inserted material explaining how and why Army Corps engineers could have missed such an egregious mistake when they conducted the E-99 Study (Sheet Pile Wall Field Load Test). Protective tarps had been placed over the test wall, hiding the mistake from the view of the engineers.

We incorporated all Dr. Rogers edits and sent the second draft back to him one week later (August 13) dutifully copying Dr. Bea. There appeared to be plenty of time to publish the paper before the ninth anniversary of the levee breaches, which was August 29, 2014. It was time to select a journal to which we would submit the article. I contacted my colleague Mark Davis, director of Tulane University's Institute for Water Resources Law & Policy. Mark recommended *Water Policy,* the official journal of the World Water Council based in Marseille, France.

Three weeks later (September 4, 2013), I wrote to Dr. Rogers: "It's been three weeks and we haven't received any edits to the second draft. So we are assuming that you feel the paper is ready for submission. We are going to accept Mark Davis's suggestion and submit it to *Water Policy* early next week. We will keep you posted. Best, Sandy."

Dr. Rogers responded: "Sandy, thanks for the update. Please keep me informed of the reviewers [sic] comments from *Water Policy*, so we can respond to those in a timely manner and keep the article on track for publication."[476]

We continued to keep Dr. Bea in the loop throughout the process. We were indebted to him for his initial candor and willingness to work with us in jumpstarting the process.

* * *

Two weeks later (September 20), H.J. uploaded the paper to the *Water Policy* website. After the submission, we noticed that the editor-in-chief of the prestigious journal was Jerome Delli Priscoli of the Army Corps, based in Arlington, Virginia.

Almost four months later (January 4,), H.J. received word that our submission had been denied. Delli Priscoli wrote, "We have now received all the comments from the reviewers, and I am sorry to tell you that your submission, 'Did the Local New Orleans Levee Board Drown its City?' has not been accepted. However, if you are willing to rewrite your submission taking into account the comments below and submit it as a new work, we will be happy to review it again."[477]

Anxiously, we reviewed the comments. Reviewer One believed that there was too much "point of view" style of wording, such as the phrase "media-powered presumptions of wrongdoing."[478] This, we felt, was a valid comment. An expert paper needed to be neutral. The reviewer also observed that the paper did not cite the IPET or the ERP report. Again, this was a valid point.

Reviewer Two was very colorful: "The paper includes a reasonable summary of the primary technical issues related to the failure of the flood protection system…but does so in an inflammatory manner that seems unnecessary to make such points."[479] Oops, we were caught again by letting our feelings show, and fortunately that was something we could fix. After all, the evidence itself was appalling enough.

The final point made by Reviewer Two was intriguing: "If the implications by the authors are correct, I encourage them to continue developing their position. Such suggested behavior should not be tolerated from the government or private industry."[480] The reviewer had noticed that we implied intentional misleading of the American public by the Army Corps. What a fascinating observation! But that was not the goal of our paper, and we would stick to the mission: to retract the wrong conclusion and replace it with the right one.

Now back to square one, H.J. and I realized that we must compromise. A prestigious journal could not tolerate innuendo, hints, or implied bad behavior.

Our first paper—the denied paper—had contained this opening sentence in the second paragraph: "Top officials with the Army Corps created a template and reported an apparently unsubstantiated story."[481] To appear neutral, we changed the sentence to this: "In this context of haste and confusion, top officials with the Army Corps and others created a template which they relayed consistently to the media and to stakeholder groups." By doing this, any possible mistake that the Army Corps spokespersons made would be blamed on the chaotic environment.

It was now clear that publishing a paper by the ninth anniversary of the 2005 flood was impossible. So, H.J. and I set our sights on the tenth anniversary and spent the extra time making the paper perfect. It turns out that despite Levees.org's ceaseless complaining about conflict of interest, the final IPET report did a satisfactory job of describing the Army Corps' levee-building mistakes. I also delved through the ERP report and the National Academy of Sciences report for citations to support the paper. There were some nice clear passages in both that described the Army Corps' mistakes. Adding citations from these three major reports improved the quality of our paper because it utilized a broad range of source data. In the end, even though Reviewers One and Two did not appear to like the paper's conclusion, they both, in their critiques, made the paper a better one.

* * *

H.J. and I polished the second draft and sent it on February 28, 2014, to Drs. Rogers and Seed for their suggestions. We wanted to make the language more neutral, and we also wanted to make it crystal clear that both the Army Corps' decision to use shorter sheet pilings and their decision to recommend the high-walls plan for the 17th Street Canal were made apparently without any recognition that risk would be significantly increased.[482] Eight weeks later (March 14), Dr. Rogers sent us his suggested edits, which we incorporated.

In high spirits, we were all ready for the second submission to *Water Policy*. But when H.J. attempted to upload the paper, he noticed that submissions are limited to 9,000 words, and our submission was 11,000 words! We had to go back to work again and cut nearly 20 percent of our beautiful paper. We proceeded to pull out everything we could without diluting the power of the paper's message. Three days later (April 5), we sent Drs. Seed and Rogers the latest edited version for their review.[483] Dr. Rogers, appreciating the need for speed, reviewed our suggested deletions and sent them back the same day.

The next day, we got a note from Dr. Seed, asking if he could have an extra week. "After a bit more than eight years, it seems that we might wait a few more days to get this re-submitted?"[484] We acquiesced.

On April 11, Dr. Seed sent his edits with this closing note: "Overall, it has become an excellent paper, and one to be proud of. Well done all!"[485]

On April 18, H.J. submitted the article—all 8,938 words—to *Water Policy*.[486]

* * *

Five months later (September 21)—an even longer wait than the first submission—H.J. received a reply from Delli Priscoli.[487] Our submission had been reviewed and if we wanted to submit a revised version, we had three weeks to do so.

Looking at the new reviewers' comments, I was alarmed. Both reviewers felt that the two endnotes on the paper should be removed. This was unacceptable to us because the endnotes addressed two common myths: the misunderstood "drive-by levee inspections" and the concept that "environmentalists had drowned New Orleans." Furthermore, unlike the first two reviewers, who had specific and valid criticisms, neither of the second two reviewers had specific suggestions. Instead, they both criticized the negative tone of the paper.

H.J. and I were, at first, incensed. We had already compromised by blaming the Army Corps' tenacity in denying its culpability for the flooding on "the context" rather than suggesting intentional disinformation. But then I understood what was happening, because I had seen this before. There was nothing wrong with the paper's tone. The reviewers did not like the paper's conclusion. And since they could not argue the merits of the paper, they tried ad hominem remarks. I was convinced that *Water Policy*, whose editor-in-chief was a member of the Army Corps, did not like the paper and had no intention of publishing it. Ever.

I communicated this thought in an email to the other authors and copied H.J.: "I believe *Water Policy* is a lost cause."[488]

I explained that I thought we should make a couple of revisions in light of some new data recently published and send our paper to the respected *Tulane Environmental Law Journal.*

Dr. Rogers responded to H.J., who promptly forwarded the email to me. "That is precisely what they want us to do. Drop it altogether."[489] He admonished us against turning to another journal. "That's one thing I've learned over the years. The publication process normally takes one to two years, and it can be very fickle."[490]

I had enormous respect for Dr. Rogers, but I had also learned to trust my intuition. I was convinced that nothing we wrote would satisfy *Water Policy*. After talking with Paul Kemp, I felt confident that we could submit our article very soon to the *Tulane Environmental Law Journal* and have it published for the tenth anniversary of the 2005 flood. On the positive side, in addition to incorporating the new data, the word limit at Tulane was 25,000 words, meaning that we could submit our longer—and better—version.

* * *

At the eleventh hour, still convinced our chances of success at *Water Policy* were nonexistent, I picked up the phone and called H.J. "If we were going to die, we would die fighting. The record will show that we submitted our paper to *Water Policy* three times, and they denied it."

H.J. wholeheartedly agreed. Just hours before the deadline on October 12, H.J. sent our third and final submission. We ignored the reviewers' recommendations. We did not remove the endnotes and we did nothing to change the tone. In fact, we even added new material we had found since our second submission. Take that, *Water Policy*!

That same day, I reached out to Tulane University. It was ten months until the tenth anniversary and time was of the essence. But, to my great dismay, the *Tulane Environmental Law Journal* was not taking any more submissions for the Summer 2015 issue. The best we could hope for, its senior editor told me, was the Winter 2015 issue. Fortunately, I got this news quickly and immediately contacted my colleague Mark Barnes for advice. Mark suggested *Louisiana History* and the *Journal of Mississippi History*. I went to their websites to find their submission requirements. Both journals had different submission rules, and the paper would need to be completely reformatted. I spent another week reformatting the footnotes and rewriting the paper more appropriately for a history journal and sent off the submissions. Then there was nothing I could do but wait.

I started to feel panicky and distraught. It began to dawn on me that I had made a huge mistake with *Water Policy*. I now understood that the process to be published in a reputable, peer-reviewed journal could take up to two years. I should have compromised as Dr. Rogers had urged us to do! Dr. Rogers was right. It was better to get "something" out there than nothing at all on time for the tenth anniversary. Attention to the 2005 flood, which took at least 1,577 lives, might not get serious attention for another five years.

While I waited for word from the two history journals, I threw myself into focusing on my family. By day, I helped Stanford and his partner move into their new apartment, and I helped Aliisa get settled into the new house that she and her husband Sam Hodges had just purchased.

Louisiana History and the *Journal of Mississippi History* wrote back, saying that it was not possible to fast-track the article for publication by August 29. Realizing that getting published in a reputable peer journal was not going to happen, I looked for other avenues. I found some credible online journals that could publish the article in our timeline. I considered publishing the paper as a white paper with editorial review. This process would be a middle ground, both academically acceptable and fast. I decided to contact the Association of State Floodplain Managers and the Natural Hazard Mitigation Association, both of which publish papers on their site. Having these fallback plans and staying busy with my family helped take the edge off my bitter disappointment over my mistake. However, I was still inconsolable, believing that I had let Levees.org's supporters down.

For two weeks, I cried myself to sleep every night.

Then, on November 4, at around eleven thirty in the morning, H.J. called me.

"*Water Policy* accepted our paper!" he said.

Annoyed, I honestly believed that H.J. was confused and had misunderstood the email from Delli Priscoli. I asked him to read the email to me, word for word.

H.J. read, "I am pleased to tell you that your submission has been accepted for publication in *Water Policy*. It was accepted on Oct 19, 2014. You will hear from IWA Publishing soon regarding the publication of your paper."[491]

After weeks of tears, I shouted over and over toward the ceiling, "I don't believe it! I don't believe it!" And I really did not believe it. After we had stood up to the reviewers and refused to change our paper, the journal had published it. I was giddy with delight. And, to this day, I am still in shock over *Water Policy*'s decision to publish our paper. And if I ever meet Delli Priscoli, I will kiss him on the lips.

* * *

It was now a glorious full ten months until the anniversary of the 2005 flood. There was plenty of time to lay out our campaign and make sure that everyone and their sister knew about the watershed journal article that soon would be published in *Water Policy*. On Thursday evening (November 13), I called Cheron Brylski and brought her up to date on the imminent *Water Policy* paper. I explained that I wanted to fly to New York City for a meeting with Dean Baquet, editor of the *New York Times*. Cheron asked me to write up a summary for her to review, and then she would get back to me the next day. I did as instructed. The next morning (November 14)

Cheron sent me an email, in a large-sized font, saying she had news from the media outlet's southeast regional reporter:

> Campbell Robertson called back right away. They want the exclusive on this. BIG TIME. He brought up [John] Schwartz. He is no longer on national desk; now at science desk. He said Schwartz would be very interested because he relied so heavily on previous WRONG opinion. Campbell wants this story BIG TIME. He's running this up the flag poll [sic] right this moment.[492]

Cheron gave me Campbell Robertson's phone number, and I called him. Robertson explained to me that, should Levees.org give the *New York Times* this exclusive, he was overwhelmingly likely to be the lead writer of the story. I had met Robertson several times before, and I was fine with this arrangement.

This turn of events was fabulous. The article would be published in a prestigious, international journal and a story about it would appear in a nationally recognized newspaper. Every media outlet in the country knew what was printed in the *New York Times*.

In my imagination, I had fantasies of friendly arguments and rollicking discussions in front of crackling fireplaces regarding levee-failure mechanisms, levee-board reform, and many other issues. Who was in charge of the infamous E-99 Study? Why was the Orleans Levee Board reorganized if the Army Corps was to blame? And how could the Army Corps have made such disastrous and egregious mistakes?

* * *

Water Policy published our paper in February 2015 titled, "Interaction between the US Army Corps of Engineers and the Orleans Levee Board preceding the drainage canal wall failures and catastrophic flooding of New Orleans in 2005."[493] We did not immediately disseminate the article because we had offered an exclusive to the *New York Times*.

After what felt like an eternity, on Sunday, May 23, Robertson made good on his word and published an excellent above-the-fold story about the *Water Policy* paper. Early in the piece, Robertson wrote:

> The article rebuts assessments of the levee system's design process that had spread responsibility around to include local officials, and it contends that fault should fall even more squarely on the corps. 'All I'm trying to do is set the record straight,' said J. David Rogers of the Missouri University of Science and Technology, the lead author of the article, whose view of the exact allocation of blame has shifted as he has come across new information.[494]

Mark Schleifstein with the *Times-Picayune* followed up the *New York Times* story with a very good article. Apparently, he had called Dr. Rogers and received additional material.

Dr. Rogers told Schleifstein that his team in 2006, "didn't have an adequate understanding of the politics behind decisions made by Congress to require increasing the heights of floodwalls along the canals instead of building gates at their mouths that would stop surge from entering."[495]

Best of all was this quote from Dr. Rogers, which was spoken less technically and more conversationally: "I could not fault the Orleans Levee Board for lobbying for the same style protection plan for the Orleans and London canals as the corps had previously recommended for the 17th Street Canal."[496] Dr. Rogers didn't blame the Orleans Levee Board for allowing the Army Corps to build the failed 17th Street Canal floodwall: "It was the corps that designed the wall and [the Orleans Levee Board] assumed it would work."[497]

* * *

The *New York Times* article was picked up by the Associated Press and reprinted in virtually every major media outlet nationwide. In addition, the story was printed in online science journals, such as *PHYS.org*[498] and *ScienceDaily*.[499]

A direct and positive result of the *Water Policy* paper—and the other articles it spawned—was the sudden cessation of any mention of Orleans Levee Board involvement in the 2005 flood. After the *New York Times* featured the journal article, coverage of the breach event became accurate, including in stories shortly afterward by John Schwarz with the *New York Times*, John Burnett[500] with NPR, and Cain Burdeau[501] with the Associated Press.

The response of other big media outlets was to make no mention of the hundreds of articles they wrote laying blame upon local levee board personnel. There were no friendly arguments taking place by firesides regarding levee-failure mechanisms and levee-board reform. There were no rollicking discussions. There was just a loud, blaring, deafening silence. But sometimes success comes in that form: silence.

H.J. submitted the *Water Policy* paper to the ASCE's Louisiana Civil Engineering Conference and Show and offered to speak on it. He was declined. The chair of the selection committee wrote saying, "The committee felt that the topics related to 2005 storm have been addressed many times at our conference and magazines. It was decided to keep more current topics."[502]

In other words, the ASCE program committee—all of them civil engineers and likely beneficiaries of the trough of work that the Army Corps made available for post-flood levee structures—did not think that this was news. This was another example of the engineering establishment taking steps to shield the Army Corps from criticism. Like Dr. Seed eight years earlier, H.J. was rebuffed from speaking at the conference because he would be presenting factual information that the engineering establishment did not want to hear.

* * *

Despite these disappointments, we considered our experiences, taken as a whole, a huge coup:

- We convinced the Berkeley team cochairs to write the retraction.
- We published the retraction in *Water Policy.*
- We obtained an exclusive story about the retraction in the *New York Times.*

And, while my dreams of nuanced, intellectual discussions over bottles of wine never materialized, I understood that I should not give up, even ten years after the anniversary of the 2005 flood. The civic activist cannot give up, even though it is tempting to do so. There were many days that I wrung my hands in confusion over the pushback. I knew that reaching success would be difficult and would take a long time. But I did not know the extent of both. Furthermore, to quit would erase everything that came before.

* * *

H.J. and I had promised Dr. Rogers that the *Water Policy* paper would be well disseminated. We kept our promise. For the next two months, I sent a personal email, including a link to both the journal article and the *New York Times* article, to every contact I had. And if I did not receive a reply, I sent it again. The responses ranged from heartwarming to aggravating. Most people, like academic experts Klaus Jacob, Mark Davis, and Scott Knowles wrote a kind, congratulatory note with assurance that they would share the links with their colleagues. Len Bahr, advisor in coastal affairs to the Louisiana governor from 1991 until his retirement in 2008, wrote a fabulous post to his intelligently written, popular blog *LaCoastPost*.[503]

We used the article's appearance in the *New York Times* as an excuse to recruit fresh blood for our letter-writing teams. On June 9, 2015, I sent an email to our supporter list. In response, more than thirty supporters signed up. At this writing, the team of letter writers is still called upon by Melissa Smith, the team leader, to correct wrong information. The *New York Times*, the *Washington Post*, NPR, and the Associated Press got the story right, but it will take years for the smaller media outlets to get on board for good.

* * *

Understanding that I would not be on this earth forever, I stayed alert for ways to memorialize the levee-breach event. So when Roy Arrigo wrote to me requesting help with verbiage for an educational plaque for his front lawn near the site of the 17th Street Canal breach, I joined forces with him. Together, we succeeded in installing a plaque vetted and fact-checked by the Louisiana SHPO in August of 2010. When Gloria Decuir contacted me about a similar plaque for her Fillmore Gardens neighborhood where she served as president, we joined forces and installed a second plaque at the London Avenue Canal breach site—also vetted and fact-checked by the SHPO—in June of 2011. In similar fashion, I joined with Linda Jackson in the Lower Ninth Ward where she served as president, and many other leaders in the neighborhood, and installed a plaque at the breach of the Industrial Canal in August of 2015. The plaques—guaranteed to last a hundred years—will tell the story for our children's children.

* * *

Organizations and agencies making bad decisions that affect millions of people should take note. All it takes is one person to unveil them. That one person might be someone they least expect. And that one person might not have specialized training or deep pockets.

I think often of the man from Alexandria—the man who said in October 2005 that the people of New Orleans deserved no help. I can still see his face—the face of the monster the Army Corps created—the face that ridiculed the people of New Orleans and blamed them for their losses. That monstrous face powered me for the next ten years and beyond to not only find the truth, but to shout it from the rooftops so that the entire nation would hear.

Epilogue

After their traumatic ordeal, Harvey and Renee Miller decided to stay permanently in Little Rock, Arkansas. They felt that, if they stayed in New Orleans, they would live in terror every time a storm formed. Besides, it made sense for them to move to where they had family, including two granddaughters.

I called Harvey in the fall of 2018. The loquacious professor was thrilled to hear from me and talked for thirty minutes about how he was glad to be near his family and of course grateful to be alive. But then his tone changed, and he confided that Little Rock just didn't "seem right." He talked about how in New Orleans there's a unique culture and, once you lived there, you became part of it. Harvey's homesickness only grew after Renee passed away on January 13, 2018.

"It's very hilly in Little Rock. In New Orleans where we lived for thirty-three years, it was as flat as a pancake. The flatness changes your perspective, because you felt like you could see forever. While we lived in New Orleans, all of our neighbors were our friends. But it's not that way in Little Rock. We know fewer people here. Maybe it's because of the hills."

List of Acronyms

ASCE: American Society of Civil Engineers

BNOBC: Bring New Orleans Back Commission

CAT: Counter Action Team

CDBG: community development block grant

CEO: chief executive officer

Citizens for 1: Citizens for 1 Greater New Orleans

CPRA: Coastal Protection and Restoration Agency

DA: district attorney

E-99 Study: Sheet Pile Wall Field Load Test

EPW: [Senate Committee on] Environment and Public Works

ERDC: Engineer Research and Development Center

ERP: External Review Panel

FEMA: Federal Emergency Management Agency

FOIA: Freedom of Information Act

GAO: Government Accountability Office (General Accounting Office prior to 2004)

GIWW: Gulf Intracoastal Waterway

IPET: Interagency Performance Evaluation Taskforce

LSU: Louisiana State University

LTE: letter to the editor

MR-GO: Mississippi River–Gult Outlet (Canal)

NHC: National Hurricane Center

NOAA: National Oceanic and Atmospheric Administration

NPR: National Public Radio

NRHP: National Register of Historic Places

NSF: National Science Foundation

NWS: National Weather Service

OPP: Outreach Process Partners

PETS: Pets Evacuation and Transportation Standards Acts

PSA: public service announcement

S&WB: [New Orleans] Sewerage & Water Board

SHPO: State Historic Preservation Office

SLAPP: strategic lawsuit against public participation

VNR: video news release

WRDA: Water Resources Development Act

Acknowledgements

Over one hundred people gave the gift of their time to the mission of Levees.org and I hope I have named them all in this book's pages. In addition, I wish to extend special thanks to the following individuals who provided interviews, material, or advice for the writing of this book:

Andy Kopplin, Ed Wikoff, Frank Donze, Gaye Tuchman, Harvey Miller, H.J. Bosworth Jr., Ivor van Heerden, Jacque Morial, Jim Huey, Jon Donley, Justin Zitler, Ken Holder, Ken McCarthy, Lee Horvitz, Mark Stein, Mary Landrieu, Mike McCrossen, Oliver Thomas, Randy Peyser, Roy Arrigo, Rudy Vorcapic, Stephen Estopinal, and Tim Doody.

About the Author

When Sandy Rosenthal is not advocating for safe levees, she plays tennis, practices yoga, and dances to Cajun music. Rosenthal is founding member of the Laurel Eagles, a mentoring program and also the Divine Revelers of Terpsichore, a Mardi Gras parading group. She has three adult children and one granddaughter. She lives in New Orleans with her husband of forty years and two small dogs.

Endnotes

1 Douglas Woolley and Leonard Shabman, "Decision-Making Chronology for the Lake Pontchartrain & Vicinity Hurricane Protection Project, Submitted to the Institute for Water Resources of the U.S. Army Corps of Engineers," March 2008, 4–25. http://levees.org/wp-content/uploads/2010/07/Woolley-Shabman-Study.pdf.

2 Ibid., 4–27.

3 C. Andersen, J. Battjes, D. Daniel, and B. Edge, "The New Orleans Hurricane Protection System: What Went Wrong and Why: A Report by the American Society of Civil Engineers External Review Panel," American Society of Civil Engineers, Reston, Virginia, 2007, ES-5. http://levees.org/wp-content/uploads/2010/06/American-Society-of-Civil-Engineers-What-Went-Wrong-ERPreport-1.pdf.

4 US Army Corps of Engineers, Public Affairs Office Honolulu Engineers District, "Honolulu District Deploys 'Power Prt' for Hurricane Katrina Relief Effort," August 30, 2005. http://levees.org/2/wp-content/uploads/2019/02/Honolulu-District-Deploys-Power-Prt-for-Hurricane-Katrina-Relief-Effort.pdf.

Chapter 1: Goodbye, New Orleans

5 John McQuaid and Mark Schleifstein, *Path of Destruction: The Devastation of New Orleans and the Coming Age of Superstorms* (New York: Little, Brown and Company, 2006), 169.

6 "Tropical Cyclone Wind Impact Product," National Oceanic and Atmospheric Administration, 2. http://www.nws.noaa.gov/images/ghls/lix/TC_WindImpact_1pager.pdf.

7 Christine Hauser and and Thomas J. Lueck, "Mandatory Evacuation Ordered for New Orleans as Storm Nears," *New York Times*, August 28, 2005.

8 Video: "Hurricane Katrina Coverage: Evacuation Ordered," CNN. August 28, 2005. https://www.youtube.com/watch?v=9upJ1vRCgoo.

9 William R. Freudenburg, et al., *Catastrophe in the Making: The Engineering of Katrina and the Disasters of Tomorrow* (Washington, DC: Island Press, 2009), 19.

10 McQuaid and Schleifstein, *Path of Destruction*, 179–180.

11 Ibid., 178.

12 Ibid., 186.

13 Ibid., 179.

14 Ibid.

15 C. Andersen, J. Battjes, D. Daniel, and B. Edge, "The New Orleans Hurricane Protection System: What Went Wrong and Why: A Report by the American Society of Civil Engineers External Review Panel," American Society of Civil Engineers, Reston, Virginia, 2007, 56. http://levees.org/wp-content/uploads/2010/06/American-Society-of-Civil-Engineers-What-Went-Wrong-ERPreport-1.pdf.

16 Flood depth map created by National Oceanic and Atmospheric Administration. http://levees.org/2/wp-content/uploads/2015/05/Untitled-1.jpg.

17 Andersen, Battjes, Daniel, Edge, "The New Orleans Hurricane Protection System: What Went Wrong and Why," 56–57.

18 Ibid.,55.

19 Ibid., 47.

20 Mike Hoss, "Firefighters share their story, video surrounding breach at 17th Street Canal," *4WWL Eyewitness Morning News*, June 20, 2006. http://levees.org/2/wp-content/uploads/2017/12/Firefighters-share-their-story-video-surrounding-breach-at-17th-Street-Canal.png.

21 Andersen, Battjes, Daniel, Edge, "The New Orleans Hurricane Protection System: What Went Wrong and Why," 55.

22 There is considerable uncertainty in the data on morbidity, particularly regarding the medical cause of death. But estimates maintain that six hundred to seven hundred victims died directly due to circumstances related to exposure to floodwaters,

approximately three hundred victims died due to other reasons relating to emergency in the region, and over 630 victims died due to circumstances related to evacuation and displacement. The latter number would continue to grow each year due to suicides. For more on the subject see, Dr. Ezra Boyd, "Fatalities Due to Hurricane Katrina's Impacts in Louisiana," Department of Geography and Anthropology. http://levees.org/wp-content/uploads/2012/11/Boyd_Dissertation.pdf.

23 McQuaid and Schleifstein, *Path of Destruction,* 200.

24 Larry King, "Tracking Hurricane Katrina," *Larry King Live*, August 29, 2005.

25 McQuaid and Schleifstein, *Path of Destruction,* 232.

26 Press Release issued by Public Affairs Office, Honolulu Engineer District US Army Corps of Engineers, "Honolulu district deploys "Power Prt" for Hurricane Katrina Relief Effort," August 30, 2005. http://levees.org/2/wp-content/uploads/2019/02/Honolulu-District-Deploys-Power-Prt-for-Hurricane-Katrina-Relief-Effort.pdf.

27 CNN staff, "Homeland Security chief defends Katrina response; Chertoff testifies as critical House report is released," *CNN Politics*, February 16, 2006.

28 McQuaid and Schleifstein, *Path of Destruction*, 246.

Chapter 2: The Flood

29 Richard D. Knabb, Jamie R. Rhome, and Daniel P. Brown, "Tropical Cyclone Report, Hurricane Katrina," National Hurricane Center, December 20, 2005, 11. http://levees.org/2/wp-content/uploads/2018/08/Tropical-Cycline-Report-Hurricane-Katrina-UPDATED-14-Sept-2011.pdf.

30 Dissertation: Dr. Ezra Boyd, "Fatalities Due to Hurricane Katrina's Impacts in Louisiana," Louisiana State University, Department of Geography and Anthropology. http://levees.org/wp-content/uploads/2012/11/Boyd_Dissertation.pdf.

31 Christopher Cooper and Robert Block, *Disaster: Hurricane Katrina and the Failure of Homeland Security* (New York: Henry Holt, 2006), 178.

32 John McQuaid and Mark Schleifstein, *Path of Destruction: The Devastation of New Orleans and the Coming Age of Superstorms* (New York: Little, Brown and Company, 2006), 250–260.

33 Ibid., 255.

34 Ibid., 287.

35 Cooper and Block, *Disaster: Hurricane Katrina and the Failure of Homeland Security*, 179.

36 McQuaid and Schleifstein, *Path of Destruction*, 299.

37 Gary Rivlin, *Katrina: After the Flood* (New York: Simon & Schuster, 2015), 129.

38 McQuaid and Schleifstein, *Path of Destruction*, 292, 298.

39 Garland Robinette, "Mayor to feds: 'Get off your asses,'" CNN, September 2, 2005.

40 Cooper and Block, *Disaster: Hurricane Katrina and the Failure of Homeland Security*, 241.

41 Ibid.

42 Sean Callebs, Sanjay Gupta, et al., "Convoys bring relief to New Orleans," CNN, September 2, 2005.

43 D'Ann Penner and Keith C. Ferdinand, *Overcoming Katrina: African American Voices from the Crescent City and Beyond* (New York: Palgrave Macmillan, 2009), xix.

44 Jarvis DeBerry, "Lt. Gen. Russel Honore showed restraint New Orleans police did not," *Times-Picayune*, August 27, 2010.

45 Jed Horne, *Breach of Faith: Hurricane Katrina and the Near Death of a Great American City* (New York: Random House, 2006).

46 Steven Horwitz, "Wal-Mart to the Rescue: Private Enterprises's Response to Hurricane Katrina," *The Independent Review*, Volume 13, Number 4, Spring 2009, 515.

47 Cooper and Block, *Disaster: Hurricane Katrina and the Failure of Homeland Security*, 177–178.

48 Jackie Loohauis-Bennett, "100 unsinkable facts about the Titanic," *Milwaukee Journal Sentinel*, April 14, 2012.

49 Mark Schleifstein, "Levee statistics point up their importance to the nation's economy," *Times-Picayune*, January 2, 2010.

50 Gail Tumulty and John R. Batty, *Voices of Angels: Disaster Lessons from Katrina Nurses* (Greta, LA: Pelican Publishing Co., 2015), 70.

51 Women of the Storm, "Katrina Facts." https://womenofthestorm.com/katrina-facts/.

52 McQuaid and Schleifstein, *Path of Destruction*, 333.

53 Ibid.

54 Ibid., 334.

55 Gary Rivlin, "Wooing Workers for New Orleans," *New York Times*, November 11, 2005.

56 Paul Purpura, "Katrina flood trial jury clears Aaron Broussard, Jefferson Parish government," *Times-Picayune*, January 27, 2016.

57 Gary Rivlin, *Katrina: After the Flood* (New York: Simon & Schuster, 2015), 86–87.

58 Ibid., 86.

59 *CNN Reports, Katrina State of Emergency* (Kansas City, MO: Andrews McMeel Publishing, 2005), 39.

60 Michael Grunwald and Susan Glasser, "Experts Say Faulty Levees Caused Much of Flooding," *Washington Post*, September 21, 2005.

61 "Minutes of the Special Board Meeting of the Board of Commissioners of the Orleans Levee District," May 30, 1991, 8. http://levees.org/2/wp-content/uploads/2018/05/May-30-1991-Special-OLB-Meeting-Minutes.pdf.

62 Grunwald and Glasser, "Experts Say Faulty Levees Caused Much of Flooding," *Washington Post*.

63 Ibid.

64 Email from Ivor van Heerden to the author, April 23, 2018. http://levees.org/2/wp-content/uploads/2019/01/Email-Ivor-van-Heerden-to-the-author-4-23-2018.pdf.

65 Ibid.

66 Rivlin, *Katrina: After the Flood*, 195.

67 Julian Borger and Jamie Wilson, "Evacuees burn to death on bus in exodus gridlock," *The Guardian*, September 25, 2005. http://levees.org/2/wp-content/uploads/2019/01/Evacuees-burn-to-death-on-bus-in-exodus-gridlock-_-World-news-_-The-Guardian.pdf.

68 Carol Christian, "8 years ago, seemingly all of Houston evacuated ahead of Hurricane Rita," *Houston Chronicle*, September 24, 2013.

69 Ralph Blumenthal, "Miles of Traffic as Texans Heed Order to Leave" *New York Times*, September 23, 2005. https://www.nytimes.com/2005/09/23/us/nationalspecial/miles-of-traffic-as-texans-heed-order-to-leave.html.

70 "Army Corps of Engineers and Hurricane Katrina," C-SPAN, September 28, 2005.

71 "Anu Mittal with GAO," C-SPAN, clip by the author, September 28, 2005.

72 Ibid.

73 Ibid.

74 W.H. Sheley Jr., "Improved Planning Needed By The Corps Of Engineers To Resolve Environmental, Technical, And Financial Issues On The Lake Pontchartrain Hurricane Protection Project," US General Accounting Office, August 17, 1982. http://levees.org/2/wp-content/uploads/2017/09/1982-GAO-report-to-Secretary-of-the-Army-.pdf.

75 Hector San Miguel, "1982 report faulted protection plan," *American Press*, September 8, 2005. http://levees.org/2/wp-content/uploads/2016/01/1982-report-faulted-protection-plan.pdf.

76 Audio recording by the author.

77 Email from Suzanne Fournier to Eugene Pawlik, Wayne Stroupe, et al, March 26, 2008. http://levees.org/2/wp-content/uploads/2017/12/Corps-response-for-media-and-casual-conversation.pdf.

78 Knabb, Rhome, and Brown, "Tropical Cyclone Report, Hurricane Katrina," National Hurricane Center, December 20, 2005, 11. http://levees.org/2/wp-content/uploads/2018/08/Tropical-Cycline-Report-Hurricane-Katrina-UPDATED-14-Sept-2011.pdf.

79 "Summary of Hearings on Hurricane Katrina (2-8-06)." http://levees.org/2/wp-content/uploads/2018/08/Summary-of-Katrina-Hearings-Compiled-by-American-Geosciences-Institute.pdf.

80 Rivlin, *Katrina: After the Flood*, 136-138.

81 Raymond Seed, ethics complaint to Dr. William Marcuson, October 30, 2007, 3. http://levees.org/2/wp-content/uploads/2014/12/WFMarcusonIII_a_-1.pdf.

82 In this book, the term "Berkeley team" has replaced the acronym "ILIT."

83 Mark Schleifstein, "Critic: Corps tried to thwart inquiry," *Times-Picayune*, November 19, 2007.

84 Seed, ethics complaint to Dr. William Marcuson.

Chapter 3: The Fairy Tale

85 Research grant W912HZ-06-1-0001, "Research and Analysis of the performance of Hurricane and Flood Protection Projects in Southeast Louisiana," issued by US Army Engineer Research and Development Center to American Society of Civil Engineers, January 31, 2006. http://levees.org/wp-content/uploads/2012/08/W912HZ-06-1-0001-Fully-Executed-2.pdf.

86 Seed, ethics complaint to Dr. William Marcuson.

87 Mike Hoss, "Firefighters share their story, video surrounding breach at 17th Street Canal," WWL-TV Channel 4, June 20, 2006.

88 Randi Rousseau, "New Orleans Firefighter: Hurricane Katrina eyewitness account made me 'Billion Dollar Man,'" WDSU-TV Channel 6, August 17, 2015.

89 Hoss, "Firefighters share their story," WWL-TV Channel 4.

90 Seed, ethics complaint to Dr. William Marcuson.

91 Ibid., 5.

92 William E. Roper, Kevin J. Weiss, and James F. Wheeler, "Water Quality Assessment and Monitoring in New Orleans Following Hurricane Katrina," George Mason University and the Environmental Protection Agency, 3. https://archive.epa.gov/emergencies/content/fss/web/pdf/roper_3.pdf.

93 Map: "Levee Breaches," created by the US Army Corps of Engineers, September 28, 2005. http://levees.org/wp-content/uploads/2011/07/LeveeBreachesMap-NO-2-1-e1309608321488.jpg.

94 Jim Tucker, "Press Release: Louisiana Republican Legislative Delegation," November 22, 2005. http://levees.org/2/wp-content/uploads/2017/04/LA-Republican-Legislative-Delegation.pdf.

95 Naomi Klein, *The Shock Doctrine: The Rise of Disaster Capitalism* (Toronto: Knopf, 2007).

96 List of Orleans Levee Board commissioners provided from the files of Wilma Heaton as Director of Governmental Affairs for Southeast Louisiana Flood Protection Authority, May 21, 2018.

97 Stephen Braun and Ralph Vartabedian, "Levee Chief Resigns Amid Nepotism Claims," *Los Angeles Times*, October 28, 2005.

98 "Head of New Orleans levee board quits," *East Valley Tribune*, October 27, 2005.

99 J. David Rogers, G. Paul Kemp, H. J. Bosworth, and Raymond B. Seed, "Interaction between the US Army Corps of Engineers and the Orleans Levee Board preceding the drainage canal wall failures and catastrophic flooding of New Orleans in 2005," *Water Policy*, August 2015, 17 (4), 707–723.

100 Public Law 109-148, extract from 3rd Supplemental, page 119 STAT.2762. http://levees.org/wp-content/uploads/2011/04/3rd-Supplemental.pdf.

101 Ibid.

102 US Army Corps of Engineers, Public Affairs Office, "Corps Points!" November 4, 2005. http://levees.org/wp-content/uploads/2011/05/Corps-Points.pdf.

103 John McQuaid and Bob Marshall, "Investigators Seeking Reasons for Levee Breaches," *Times-Picayune*, October 2, 2005. https://rense.com/general67/investigatorsseeking.htm.

104 Amy Liu, Bruce Katz, Matt Fellowes, and Mia Mabanta, "Housing Families Displaced by Katrina: A Review of the Federal Response to Date," Brookings, November 1, 2005. https://www.brookings.edu/research/housing-families-displaced-by-katrina-a-review-of-the-federal-response-to-date/.

Chapter 4: The Face of the Monster

105 Amy Liu, Bruce Katz, Matt Fellowes, and Mia Mabanta, "Housing Families Displaced by Katrina: A Review of the Federal Response to Date," Brookings, November 1, 2005. https://www.brookings.edu/research/housing-families-displaced-by-katrina-a-review-of-the-federal-response-to-date/.

106 Ibid.

107 Ibid.

108 US Army Corps of Engineers, Public Affairs Office, "Corps Points!" November 4, 2005. http://levees.org/wp-content/uploads/2011/05/Corps-Points.pdf.

109 American Geosciences Institute, "Geoscience Policy." http://www.agiweb.org/gap/legis109/katrina_hearings.html#nov02b.

110 *CNN Reports, Katrina State of Emergency* (Kansas City, MO: Andrews McMeel Publishing, 2005), 39.

111 Raymond Seed, ethics complaint to William Marcuson, October 30, 2007, 7. http://levees.org/2/wp-content/uploads/2014/12/WFMarcusonIII_a_-1.pdf.

112 "Louisiana Legislature Log." http://laleglog.blogspot.com/2005_11_01_archive.html.

113 Anu Mittal, "GAO Testimony: Lake Pontchartrain and Vicinity Hurricane Protection Project," September 28, 2005.

114 Brett Martel, "Investigations into La. Levee Breaks Mount," Associated Press, November 10, 2005.

115 Ibid.

116 Jim Shannon, "Sen Boasso Discusses His Levee Consolidation Proposal," WAFB9 TV, August 24, 2006. http://www.wafb.com/story/5319652/sen-boasso-discusses-his-levee-consolidation-proposal/.

117 Blog post: Jeffrey Sadow, "Louisiana Legislature Log," November 17, 2005. http://levees.org/2/wp-content/uploads/2019/04/Jeffrey-Sadow-blog-post-Nov-17-2005.pdf.

118 Ibid.

119 Ibid.

120 Ibid.

121 Business Council of New Orleans and the River Region, "It's Time for Experts to Manage Levee Safety." http://levees.org/2/wp-content/uploads/2017/04/Ad-in-Times-Picayune-11-16-05.pdf.

122 Video: "John Georges for One Levee Board." https://www.youtube.com/watch?v=v2n4_qz-prg.

123 Public Law 109-148, extract from 3rd Supplemental, page 119 STAT.2762. http://levees.org/wp-content/uploads/2011/04/3rd-Supplemental.pdf.

124 Official Journal of the House of Representatives of the State of Louisiana, November 20, 2005. http://house.louisiana.gov/H_Journals/H_Journals_All/2005_1stESJournals/051ES%20-%20HJ%201120%2012.pdf.

125 Levees.org original homepage. http://levees.org/2/wp-content/uploads/2017/10/Original-Levees.org-homepage.png.

126 Email from Dr. Michele C. Spector to patient Sandy Rosenthal: "The medical description of your hearing loss is a normal sloping to moderate high frequency sensorineural hearing loss in the left ear, with a severe noise notch at 3000 Hz and a normal sloping to severe high frequency sensorineural hearing loss in the right, with a profound noise notch at 3000 Hz. You can characterize yourself as severely hearing impaired." http://levees.org/2/wp-content/uploads/2019/01/Email-Michele-Spector-to-the-author.pdf.

127 Bruce Alpert, "Bush '05: 'No way to imagine America without New Orleans,'" *Times-Picayune*, August 28, 2015.

128 Frank Donze, "Failure of levee merger sparks outrage—Supporters say they won't let issue drop," *Times-Picayune*, November 22, 2005.

129 Bring New Orleans Back, "Infrastructure Committee; Levees and Flood Protection Sub-Committee." http://levees.org/2/wp-content/uploads/2017/10/Bring-New-Orleans-Back-Infrastructure-Committee-Levees-and-Flood-Protection-Sub-Committee.pdf.

130 Jim Tucker, "Press Release: Louisiana Republican Legislative Delegation," November 22, 2005. http://levees.org/2/wp-content/uploads/2017/04/LA-Republican-Legislative-Delegation.pdf.

131 Donze, "Failure of levee merger sparks outrage" *Times-Picayune.*

132 Email from Oliver Thomas to the author, January 10, 2016. http://levees.org/2/wp-content/uploads/2019/01/Email-Oliver-Thomas-to-the-author.pdf.

133 Donze, "Failure of levee merger sparks outrage" *Times-Picayune.*

134 Jacques Morial shared this information in an interview on December 3, 2015.

135 Haynie & Associates has listed Citizensfor1GreaterNewOrleans on its website as a customer. This information is no longer available on the internet.

136 BGR, "Backers of bloated assessor system try new play," *Lafayette Advisor*, April 20, 2006. http://levees.org/2/wp-content/uploads/2018/02/BGR-›-In-The-News-›-Backers-of-bloated-assessor-system-try-new-play.pdf.

137 Charles C. Mann, "The long, strange resurrection of New Orleans," *Fortune Magazine*, August 29, 2006.

138 Blog post: Seymour D. Fair, "Levee Board, One Voice/Signs, Signs, Everywhere are Signs," December 12, 2005. http://thethirdbattleofneworleans.blogspot.com/2005/12/levee-board-one-voicesigns-signs.html.

139 Public Affairs Research Council of Louisiana, "PAR Says Consolidate New Orleans Assessors." http://parlouisiana.org/par-says-consolidate-new-orleans-assessors/.

140 Photo: "Chris Granger with Times Picayune interviews Mary Burns." https://www.flickr.com/photos/24669697@N05/6274215326/in/photolist-ayqZCj-6Wor9F-9wTnPP-9wWmUU.

141 Photo: "Yard sign at corner of St. Charles Avenue and Soniat Street." http://levees.org/2/wp-content/uploads/2017/10/One-Levee-Board-Yardsign.jpg.

142 Hearing before the Committee on Environment and Public Works, "Evaluate the Degree to which the Preliminary Findings on the Failure of the Levees are Being Incorporated into the Restoration of the Hurricane Protection," November 17, 2005. https://www.govinfo.gov/content/pkg/CHRG-109shrg39525/pdf/CHRG-109shrg39525.pdf.

143 Ibid.

144 Transcript: "Hearing before the Committee on Homeland Security and Governmental Affairs, United States Senate," December 15, 2005. https://www.govinfo.gov/content/pkg/CHRG-109shrg26746/html/CHRG-109shrg26746.htm.

145 Video: Hearing before the Committee on Homeland Security and Governmental Affairs, US Senate, December 15, 2005. http://www.c-span.org/video/?190352-1/maintaining-new-orleans-levees.

146 Transcript: "Hearing before the Committee on Homeland Security and Governmental Affairs, United States Senate," December 15, 2005. https://www.govinfo.gov/content/pkg/CHRG-109shrg26746/html/CHRG-109shrg26746.htm.

147 Gordon Russell, "Levee inspections only scratch the surface," *Times-Picayune*, November 25, 2005.

148 Ibid.

149 Bob Marshall is no relation to Nancy Marshall, who pushed for the property-evaluation assessor office consolidation.

150 Bob Marshall, "N.O. levee inspections fell short of federal mandate," *Times-Picayune*, December 5, 2005.

151 Hearing before the Committee on Homeland Security and Governmental Affairs, US Senate, December 15, 2005. http://www.c-span.org/video/?190352-1/maintaining-new-orleans-levees.

152 Ibid.

153 Ibid.

154 Ibid.

155 Ibid.

156 Ibid.

157 Ralph Vartabedian and Stephen Braun, "Fatal Flaws: Why the Walls Tumbled in New Orleans," *Los Angeles Times*, January 17, 2006.

158 Ibid.

159 Editorial, "Death of an American City," *New York Times*, December 11, 2005.

160 William Branigin, "Senate Blocks Arctic Drilling Provision," *Washington Post*, December 21, 2005.

161 Ibid.

162 109th Congress Public Law 148, Department of Defense, Emergency Supplemental Appropriations to Address Hurricanes in the Gulf of Mexico. https://www.govinfo.gov/content/pkg/PLAW-109publ148/html/PLAW-109publ148.htm.

163 Louisiana Recovery Authority, October 17, 2005. http://www.lra.louisiana.gov/index.cfm?md=pagebuilder&tmp=home&pid=5.

164 Mitch Landrieu, *In the Shadow of Statues: A White Southerner Confronts History* (New York: Viking, 2018), 117.

165 Email from Lauren Solis to the author, January 3, 2006. http://levees.org/2/wp-content/uploads/2019/01/Email-Lauren-Solis-to-author-1-3-2006.pdf.

166 LinkedIn screenshot for Lauren Solis: http://levees.org/2/wp-content/uploads/2014/12/Screen-shot-2011-04-24-at-3.14.08-PM-2.png.

167 "New Orleans Pumped Out, Army Says," Associated Press, October 12, 2005.

168 Spencer S. Hsu, "Bush Adviser Acknowledges Lack of Preparation for Katrina," *Washington Post*, October 22, 2005.

169 Email from Steve Rosenthal to Lauren Solis, January 3, 2006. http://levees.org/2/wp-content/uploads/2019/01/Email-from-Stephen-Rosenthal-to-Lauren-Solis.pdf.

170 LinkedIn screenshot for Lauren Solis: http://levees.org/2/wp-content/uploads/2014/12/Screen-shot-2011-04-24-at-3.14.08-PM-2.png.

Chapter 5: The Force to Be Reckoned With

171 109th Congress Public Law 148, Department of Defense, "Emergency Supplemental Appropriations to Address Hurricanes in the Gulf of Mexico." https://www.govinfo.gov/content/pkg/PLAW-109publ148/html/PLAW-109publ148.htm.

172 Mitch Landrieu, *In the Shadow of Statues: A White Southerner Confronts History* (New York: Viking, 2018), 118–119.

173 Email from Mary Burns to the author, January 4, 2006. http://levees.org/2/wp-content/uploads/2019/01/Email-Mary-Burns-to-the-author.pdf.

174 Blog Post: Sandy Rosenthal's kickoff rally day statement. http://levees.org/2006/01/21/founders-rally-day-speech/.

175 Allen Johnson Jr., "Protesters: Hold corps responsible for N.O. flood," *The Advocate*, January 22, 2006.

176 Ibid.

177 Raymond Seed, ethics complaint to Dr. William Marcuson, October 30, 2007, 12. http://levees.org/2/wp-content/uploads/2014/12/WFMarcusonIII_a_-1.pdf.

178 Department of the Army, US Army Engineer Research and Development Center, "Research Grant Schedule." http://levees.org/2/wp-content/uploads/2017/09/Response-to-FOIA-Request-Grant-from-USACE-to-ASCE-Jan-31-2006.pdf.

179 Email from Robert Bea to the author, January 31, 2006. http://levees.org/2/wp-content/uploads/2019/01/Email-Robert-Bea-to-author-1-31-2006.pdf.

180 Email from Travis L. Constanza to the author, February 6, 2006. http://levees.org/2/wp-content/uploads/2019/01/Email-from-Travis-Costanza-to-author-2-6-2006.pdf.

181 Ibid.

182 Ibid.

183 Ibid.

184 Ibid.

185 Email from David Schad to the author, February 7, 2006. http://levees.org/2/wp-content/uploads/2019/01/Email-from-David-Schad-to-author-2-7-2006.pdf.

186 Letter from the author to Stephen Sabludowsky, February 18, 2006. http://levees.org/2/wp-content/uploads/2019/01/Press-exclusive-to-Bayou-Buzz-2-18-2006.pdf.

187 Email from Andy Kopplin to the author, November 19, 2017. http://levees.org/2/wp-content/uploads/2019/01/Email-from-Andy-Kopplin-to-author-11-19-2017.pdf.

188 Email from Andy Kopplin to the author, November 17, 2017. http://levees.org/2/wp-content/uploads/2019/01/Email-Andy-Kopplin-to-author-11-17-2017.pdf.

189 Ibid.

190 Office of the Governor, "Gov. Blanco's speech to the Louisiana Legislature," February 6, 2006. http://levees.org/2/wp-content/uploads/2017/11/Kathleen-Blanco-speech-Feb-6-2006.pdf.

191 Ibid.

192 "CF1 Rally in Baton Rouge." http://levees.org/2/wp-content/uploads/2017/02/Email-urging-supporters-to-attend-BR-rally-.png.

193 Office of the Governor, "Gov. Blanco's speech to the Louisiana Legislature," February 6, 2006. http://levees.org/2/wp-content/uploads/2017/11/Kathleen-Blanco-speech-Feb-6-2006.pdf.

194 Ibid.

195 Ibid.

196 Public Law 109-148, extract from 3rd Supplemental, page 119 STAT.2762. http://levees.org/wp-content/uploads/2011/04/3rd-Supplemental.pdf.

197 109th Congress Public Law 148. Department of Defense, Emergency Supplemental Appropriations to Address Hurricanes in the Gulf of Mexico, and Pandemic Influenza Act, 2006. https://www.gpo.gov/fdsys/pkg/PLAW-109publ148/html/PLAW-109publ148.htm.

198 Blog post: *Between The Lines*, "Democrats prefer preservation over responsibility in session." http://jeffsadow.blogspot.com/2006_02_12_archive.html?m=1.

199 P. J. Huffstutter and Sam Quinones, "Merger of Louisiana Levee Boards, OKd," *Los Angeles Times*, February 17, 2006.

200 Email from Roy Arrigo to the author, December 26, 2018. http://levees.org/2/wp-content/uploads/2019/01/Email-Roy-Arrigo-to-author-12-26-2019.pdf.

201 Huffstutter and Quinones, "Merger of Louisiana Levee Boards, OKd," *Los Angeles Times*.

202 "ACT No. 1, Senate Bill No. 8," First Extraordinary Session, 2006. http://levees.org/2/wp-content/uploads/2017/12/First-Extraordinary-Session-2006-Senate-Bill-No-8.pdf.

203 "Senate Bill No. 95," First Extraordinary Session, 2005. http://www.legis.la.gov/Legis/ViewDocument.aspx?d=326449.

204 Office of the Governor, "Gov. Blanco's speech to the Louisiana Legislature," February 6, 2006. http://levees.org/2/wp-content/uploads/2017/11/Kathleen-Blanco-speech-Feb-6-2006.pdf.

205 Huffstutter and Quinones, "Merger of Louisiana Levee Boards, OKd," *Los Angeles Times*.

206 Ibid.

207 Richard Cowan and Caren Bohan, "U.S. House panel approves new hurricane aid," *Reuters*, March 8, 2006.

208 Jim Moran, "Much Left To Do Post Katrina," *Herndon Connection*, March 17, 2006.

209 Email from John Schwartz to the author, October 31, 2006. http://levees.org/2/wp-content/uploads/2018/12/Email-string-John-Schwartz.pdf.

210 Seed, ethics complaint to Dr. William Marcuson.

211 Email from Garret Graves to the author, April 17, 2006. http://levees.org/2/wp-content/uploads/2019/01/Email-Garret-Graves-to-author-4-17-2006.pdf.

212 John Schwartz, "Hurricane Protection Plan Flawed, Engineers Say," *New York Times*, May 2, 2006.

213 Ibid.

214 Ibid.

215 Email from Robert Bea to the author, May 9, 2006. http://levees.org/2/wp-content/uploads/2019/01/Email-Robert-Bea-to-author-5-9-2006.pdf.

216 Seed, ethics complaint to Dr. William Marcuson.

217 Ibid.

218 Melissa McNamara, "N.O. Levees Swamped With Criticism," CBS News/Associated Press, May 22, 2006.

219 Ibid.

220 Ibid.

221 Editorial: "In Katrina disaster, human error claimed heavy toll," *USA Today*, May 23, 2006.

222 Mark Schleifstein, "Study: Corps decisions, not Orleans Levee Board, doomed canal walls in Katrina," *Times-Picayune*, August 7, 2015.

223 Email from Robert Verchick to Oliver Houck and the author, February 2, 2009. http://levees.org/2/wp-content/uploads/2019/01/Email-to-Oliver-Houck-and-author-2-2-2009.pdf.

224 John Schwartz, "Army Corps Admits Flaws in New Orleans Levees," *New York Times*, June 1, 2006.

225 Ibid.

226 John McQuaid and Mark Schleifstein, "Washing Away," five-part series, *Times-Picayune*, June 23–27, 2002.

227 Society of Environmental Journalists annual conference in New Orleans, September 4, 2014.

228 Schwartz, "Army Corps Admits Flaws in New Orleans Levees," *New York Times*.

229 Ibid.

230 Ibid.

231 John Schwartz, "Too Bad Hippocrates Wasn't an Engineer," *New York Times*, June 11, 2006.

232 Betty Ann Bowser, "Costly New Orleans Levee Repairs May Be Inadequate," *PBS NewsHour*, June 12, 2006.

233 C. Andersen, J. Battjes, D. Daniel, and B. Edge, "The New Orleans Hurricane Protection System: What Went Wrong and Why: A Report by the American Society of Civil Engineers External Review Panel," American Society of Civil Engineers, Reston, Virginia, 2007, 39. http://levees.org/wp-content/uploads/2010/06/American-Society-of-Civil-Engineers-What-Went-Wrong-ERPreport-1.pdf.

234 Ibid., 31.

235 Seed, ethics complaint to Dr. William Marcuson.

236 Ibid.

237 Email from Andy Kopplin to the author, November 17, 2017. http://levees.org/2/wp-content/uploads/2019/01/Email-Andy-Kopplin-to-author-11-17-2017.pdf.

238 Ibid.

239 Email from Annette Sisco to the author, June 21, 2006. http://levees.org/2/wp-content/uploads/2018/05/Email-string-Sisco-and-Rosenthal-June-2006.pdf.

240 David Vitter, "Moving Forward after Hurricanes Katrina and Rita," Hearing before the Committee on Environment and Public Works, US Senate, New Orleans, February 26, 2007.

241 Email from Adam Sharp forwarded to the author, August 14, 2006. http://levees.org/2/wp-content/uploads/2019/01/Email-from-Adam-Sharp-forwarded-to-author-8-14-2006.pdf.

242 Seed, ethics complaint to Dr. William Marcuson.

243 Ibid.

244 Email from Michael Grunwald to the author on August 30, 2006. http://levees.org/2/wp-content/uploads/2019/01/email-chain-Michael-Grunwald-and-Sandy-Rosenthal.pdf.

245 Ibid.

246 Screenshot: Citizensfor1.com. http://levees.org/2/wp-content/uploads/2017/11/Screen-Shot-2017-11-01-at-11.58.51-AM.png.

247 Email from Citizensfor1GreaterNewOrleans.com to the author, August 23, 2006. http://levees.org/2/wp-content/uploads/2019/01/eBlast-to-author-8-23-2006.pdf.

248 Doug Simpson, "La. Voters to Weigh Fate of Levee Boards," Associated Press, September 29, 2006.

249 "Louisiana lauded for voting to scrap its levee system," Associated Press, October 2, 2006.

250 Op-ed article: Sandy Rosenthal, "Congress must look into failure of levees," *Times-Picayune*, November 20, 2006.

251 Email from Stanford Rosenthal to the author, December 18, 2006. http://levees.org/2/wp-content/uploads/2018/05/Stanford-Rosenthal-UASCE-is-watching-us.pdf.

Chapter 6: Figuring Out the Allies

252 Email from jr7402@bellsouth.net to the author, January 4, 2007. http://levees.org/2/wp-content/uploads/2019/01/Email-from-anonymous-to-author-1-4-2007.pdf.

253 Cain Burdeau, "Group Wants 9-11-Style Panel on Levees," Associated Press, February 5, 2007.

254 Ibid.

255 "Allied Organizations." http://levees.org/other/allied-organizations/.

256 Gary Scheets, "Memo: FEMA waived own advice," *Times-Picayune*, April 25, 2007.

257 Email from Ed Wikoff to the author, November 27, 2017. http://levees.org/2/wp-content/uploads/2019/01/Email-Ed-Wikoff-to-author-11-27-2017.pdf.

258 Jarvis Deberry, "Road Home has callously disregarded Louisianans—from its start till now," *Times-Picayune*, March 22, 2014.

259 Editorial: "Unmatched Destruction," *New York Times*, February 13, 2007.

260 Ibid.

261 Ken McCarthy, "It was the levees, stupid," *FoodMusicJustice.org*, January 2007. Blog no longer active.

262 This new link to the blog post is also no longer functional.

263 Email from Ken McCarthy to the author, January 12, 2007. http://levees.org/2/wp-content/uploads/2019/01/Email-Ken-McCarthy-to-author-1-12-2007.pdf.

264 Video: "Levees.org PSA on nationwide protection," July 29, 2008. https://www.youtube.com/watch?v=swJ9_2Adl9A.

265 Press release by The Brylski Company, May 16, 2007. http://levees.org/2/wp-content/uploads/2019/01/Press-release-by-Brylski-Company-5-16-2007.pdf.

266 Email from Pat Doherty to the author, May 10, 2007. http://levees.org/2/wp-content/uploads/2019/01/Email-Pat-Doherty-to-author-5-10-2007.pdf.

267 Email from Erin Carlson to the author, May 11, 2007. http://levees.org/2/wp-content/uploads/2019/01/Email-Erin-Carlson-to-author-5-11-2007.pdf.

268 Federal Contracts for Federal Fiscal Year 2007. http://levees.org/2/wp-content/uploads/2018/05/Federal-Contracts-for-Federal-Fiscal-Year-2007-State-of-Louisiana.pdf.

269 Spreadsheet: Funds paid by Army Corps of Engineers to Outreach Process Partners from 2007 to 2015. http://levees.org/2/wp-content/uploads/2018/12/Funds-paid-by-USACE-to-OPP-2007-to-2015.pdf.

270 Press release issued by The Brylski Co., May 16, 2007. http://levees.org/2/wp-content/uploads/2019/01/Press-release-by-Brylski-Company-5-16-2007.pdf.

271 Legislation in 110th Congress 1st Session to establish the 8/29 Commission. http://levees.org/2/wp-content/uploads/2015/02/829-Commission.pdf.

272 Senator Mary Landrieu, 110th Congress (2007–2008) S.2826 – 8/29 Investigation Team Act. http://levees.org/2/wp-content/uploads/2017/11/S.2826-110th-Congress-2007-2008_-8_29-Investigation-Team-Act-_-Congress.pdf.

273 Seed, ethics complaint to Dr. William Marcuson.

274 Press release issued by the American Society of Civil Engineers, June 1, 2007. http://levees.org/wp-content/uploads/2010/11/ASCE_Press_Release_June.pdf.

275 Ibid.

276 Ibid.

277 Email from Gordon Boutwell to Robert Bea forwarded to the author, June 7, 2007. http://levees.org/2/wp-content/uploads/2019/01/Email-from-Gordon-Boutwell-to-Robert-Bea-forwarded-to-author-6-9-2007.pdf.

278 Email from Larry Roth to Stephan Butler, June 21, 2007. http://levees.org/2/wp-content/uploads/2019/01/Email-from-Larry-Roth-to-Stephen-Butler-7-21-2007-.pdf.

279 Screenshot: "Katrina Study Findings Presentations," American Society of Civil Engineers, recorded on January 12, 2012. http://levees.org/2/wp-content/uploads/2017/09/List-of-ASCE-presentations-.png.

280 Video: Larry Roth, excerpts from "The New Orleans Levees: The Worst Engineering Catastrophe in U.S. History," American Society of Civil Engineers. https://www.youtube.com/watch?v=prKJcog20-Y.

281 Video: "Elite Engineering Group Does PR Show to Protect Corps of Engineers' Reputation." https://www.youtube.com/watch?v=lLNtA-5C5m0&t=28s.

282 Screenshot: "Katrina Study Findings Presentations," American Society of Civil Engineers, recorded on January 12, 2012. http://levees.org/2/wp-content/uploads/2017/09/List-of-ASCE-presentations-.png.

283 Email from Gordon Boutwell to the author, June 20, 2007. http://levees.org/2/wp-content/uploads/2019/01/Email-Gordon-Boutwell-to-author-6-20-2007.pdf.

284 Blog post: "Corps of Engineers is culprit not savior," August 26, 2007. http://levees.org/2007/08/26/corps-of-engineers-is-culprit-not-savior/.

285 Video: "Elite Engineering Group Does PR Show to Protect Corps of Engineers' Reputation." https://www.youtube.com/watch?v=lLNtA-5C5m0.

286 Screenshot: American Society of Civil Engineers, "Katrina Study Findings Presentations." http://levees.org/wp-content/uploads/2012/01/Screen-shot-2012-01-12-at-4.42.52-PM.png.

287 List of IPET leadership. http://levees.org/2/wp-content/uploads/2018/12/IPET-Leadership.pdf.

288 Video: "New Orleans Levee Spin 101," Levees.org, November 5, 2007. https://www.youtube.com/watch?v=XauhgHNgPw0.

289 Email from the author to Cheron Brylski, November 7, 2007. http://levees.org/2/wp-content/uploads/2019/01/Email-from-author-to-Cheron-Brylski-11-7-2007.pdf.

290 Letter from Tom Smith to the author, November 10, 2007. http://levees.org/wp-content/uploads/2010/11/Rosethal.ltr_.pdf.

291 Email from Paul Harrison to the author, November 11, 2007. http://levees.org/2/wp-content/uploads/2019/01/Email-from-Paul-Harrison-to-author-11-11-2007.pdf.

292 Email from Tom Smith to the author, November 13, 2007. http://levees.org/2/wp-content/uploads/2019/01/Email-Tom-Smith-to-author-11-13-2007.pdf.

293 Email author to Tom Smith, November 13, 2007. http://levees.org/2/wp-content/uploads/2019/01/Email-author-to-Tom-Smith-11-13-2007.pdf.

294 Dan Shea, "Controversial Levees.org video," *Times-Picayune*, November 14, 2007.

295 Blog post: Sandy Rosenthal, "Army Corps of Engineers caught using online warfare to attack Levees.org." Updated December 1, 2018. http://levees.org/personal-attacks-and-lies-posted-to-internet-by-army-corps-of-engineers/.

296 Email from the author to endorsers of the 8/29 Investigation, November 16, 2007. http://levees.org/2/wp-content/uploads/2019/01/Email-from-author-to-Endorsers-of-820-Investigation-11-16-2007.pdf.

297 Ibid.

298 Mark Schleifstein, "Critic: Corps tried to thwart inquiry," *Times-Picayune*, November 19, 2007.

299 Ibid.

300 Email from Samantha Everett to the author, November 18, 2007. http://levees.org/2/wp-content/uploads/2019/01/Email-Samantha-Everett-to-author-11-18-2007.pdf.

301 Email from the author to Samantha Everett, November 22, 2007. http://levees.org/2/wp-content/uploads/2019/01/Email-author-to-Samantha-Everett-11-22-2007.pdf.

302 Letter from Samantha Everett to Tom Smith, December 12, 2007. http://levees.org/wp-content/uploads/2010/11/Cooley-letter-to-ASCE.pdf.

303 Email from Lu Christi to USACE DC HQ, December 12, 2007. http://levees.org/2/wp-content/uploads/2018/12/Emails-in-DC-USACE-HQ-on-reposting-satire-to-YouTube-December-2007-1.pdf.

304 Email from Steven Wright to Eugene Pawlik, December 13, 2007. http://levees.org/2/wp-content/uploads/2018/12/Emails-in-DC-USACE-HQ-on-reposting-satire-to-YouTube-December-2007-1.pdf.

305 Email from Eugene Pawlik to Joan Burhman, December 12, 2007. http://levees.org/2/wp-content/uploads/2018/12/Emails-in-DC-USACE-HQ-on-reposting-satire-to-YouTube-December-2007-1.pdf.

306 Eugene (Gene) Pawlik's LinkedIn profile states that he is currently Supervisory Public Affairs Specialist at the US Army Corps of Engineers.

307 Email from Eugene Pawlik to Joan Buhrman, December 14, 2007. http://levees.org/2/wp-content/uploads/2018/12/Emails-in-DC-USACE-HQ-on-reposting-satire-to-YouTube-December-2007-1.pdf.

308 Ibid.

309 Ibid.

310 Email from Stanford Rosenthal to Matt Faust, August 12, 2008. http://levees.org/2/wp-content/uploads/2019/01/Email-Stanford-Rosenthal-to-Matt-Faust-8-12-2008.pdf.

311 Email from Geralyn M. Ryan to other Public Affairs officers in New Orleans, December 21, 2007. http://levees.org/2/wp-content/uploads/2018/12/Corps-PA-office-observes-home-video-12-21-2007.pdf.

Chapter 7: Foiling the Ruse

312 Adam Nossiter, "In Court Ruling on Floods, More Pain for New Orleans," *New York Times*, February 1, 2008.

313 Judge Stanwood Duval, United States District Court Eastern District of Louisiana, Civil Action No. 05-4182, January 30, 2008, 44. http://levees.org/2/wp-content/uploads/2018/12/Dismissal.pdf.

314 Nossiter, "In Court Ruling on Floods, More Pain for New Orleans," *New York Times*.

315 D. D. Gaillard, "The Washington Aqueduct and Cabin John Bridge," National Geographic Magazine, Volume VIII, Number 12, December 1897.

316 Gary Rivlin, *Katrina: After the Flood* (New York: Simon & Schuster, 2015), 92.

317 Blog post: Sandy Rosenthal, "Chertoff defends Corps of Engineers in New Orleans flooding." http://levees.org/2008/03/02/chertoff-defends-corps-of-engineers-in-new-orleans-flooding/.

318 Erik Sofge, "13 Tough Questions for the Army Corps of Engineers' Flood Reconstruction Chief," *Popular Mechanics*, June 24, 2008.

319 Email from Ken McCarthy to the author, February 29, 2008. http://levees.org/2/wp-content/uploads/2019/01/Email-Ken-McCarthy-to-author-2-29-2008.pdf.

320 Email from Dr. Bob Bea to the author, December 29, 2007. http://levees.org/2/wp-content/uploads/2019/01/Email-Robert-Beat-to-author-12-29-2007.pdf.

321 Flood depth map created by National Oceanic and Atmospheric Administration. http://levees.org/2/wp-content/uploads/2015/05/Untitled-1.jpg.

322 Email from Kathy Gibbs to Suzanne Fournier and Eugene Pawlick, March 26, 2009. http://levees.org/2/wp-content/uploads/2017/11/PR-SHOW-email-chain.pdf.

323 Email from Suzanne Fournier to Eugene Pawlik and Wayne Stroupe, March 26, 2008. http://levees.org/2/wp-content/uploads/2017/12/Corps-response-for-media-and-casual-conversation.pdf.

324 Email from Wayne Stroupe to Denise Frederick, March 26, 2008. http://levees.org/2/wp-content/uploads/2019/01/Email-Wayne-Stroupe-to-Denise-Frederick-3-26-2009.pdf.

325 Email from Wade Habshey to Rene Poche, March 26, 2008. http://levees.org/2/wp-content/uploads/2018/06/Email-Corps-dispatched-two-personnel.pdf.

326 Email from Wayne Stroup to New Orleans District Public Affairs officers, June 3, 2008. http://levees.org/2/wp-content/uploads/2017/12/ERDC-notifies-N.O.-District-.pdf.

327 Email from E. R. DesOrmeaux to Louisiana ASCE members, April 2, 2008. http://levees.org/2/wp-content/uploads/2019/01/Email-E.R.-DesOrmeaux-to-ASCE-Louisiana-members-4-2-2008.pdf.

328 The Association of Environmental and Engineering Geologists, the Association of State Floodplain Managers, the American Society of Civil Engineers, and the Hazard Caucus Alliance, "Levee Protection: Working with the Geology and Environment to Build Resiliency,"

Washington, DC, June 19, 2008. http://levees.org/2/wp-content/uploads/2017/12/leveebriefing-flyer-06081.pdf.

329 Ibid.

330 Ed Vogel, "Levee Break Floods Fernley," *Las Vegas Review-Journal*, January 6, 2008.

331 William Petroski, "River engulfs parts of Cedar Rapids," *Des Moines Register*, June 13, 2008.

332 George Hesselberg and Ron Seely, "Lake Delton vanishes into Wisconsin River," *The Chippewa Herald*, June 10, 2008.

333 The Association of Environmental and Engineering Geologists, the Association of State Floodplain Managers, the American Society of Civil Engineers, and the Hazard Caucus Alliance, "Levee Protection: Working with the Geology and Environment to Build Resiliency."

334 Keith O'Brien and Stephen Smith, "New Orleans passes test, faces challenge," *Boston Globe*, September 3, 2008.

335 Mark Schleifstein, "American Society of Civil Engineers finds no ethical violations in its own Katrina levee review," *Times-Picayune*, April 6, 2009.

336 Press release: "Engineering society was a whitewash says Levees.org," The Brylski Company, April 6, 2009.

337 Email from Gordon Boutwell to the author, June 20, 2007. http://levees.org/2/wp-content/uploads/2019/01/Email-Gordon-Boutwell-to-author-6-20-2007.pdf.

338 Email from Karen Collins to Rene Poche, April 8, 2009. http://levees.org/2/wp-content/uploads/2017/12/Corps-observed-ASCE-press-release-retracted.pdf.

339 8/29 Investigation webpage. http://levees.org/investigation-home/.

340 Letter from Sherwood Boehlert to David Mongan, September 12, 2008. http://levees.org/2/wp-content/uploads/2015/04/BoehlertReport.pdf.

341 Mark Schleifstein, "Disaster-investigation rules changed by American Society of Civil Engineers," *Times-Picayune*, November 24, 2009.

342 Letter from Anthony Stiegler to the author, November 19, 2010. http://levees.org/wp-content/uploads/2010/11/Cooley.pdf.

343 Comment posted by stevonawlins, "See you at the New Orleans Institute," *Times-Picayune*, October 24, 2008.

344 Letter from John R. Crane to Mary L. Landrieu, September 17, 2009. http://levees.org/wp-content/uploads/2009/09/2009_us_IG-Response-to-Landrieu2.pdf.

345 *Corbett Report*, "Episode 332—The Weaponization of Social Media," March 2, 2018.

346 Excerpt of manuscript chapter edited by Justin Zitler and sent to the author, February 3, 2018. http://levees.org/2/wp-content/uploads/2019/01/Edited-Chapter-8-by-Justin-Zitler.pdf.

347 Video: "Corps of Engineers caught harassing citizens on Internet," *Nightwatch*, WWL-TV Channel 4, December 25, 2008. https://www.youtube.com/watch?v=jDwuMBOPrrQ.

348 Ibid.

349 Ibid.

350 Letter from John R. Crane to Mary L. Landrieu, September 17, 2009. http://levees.org/wp-content/uploads/2009/09/2009_us_IG-Response-to-Landrieu2.pdf.

351 Letter from Col. Alvin B. Lee to the author, December 18, 2008. http://levees.org/wp-content/uploads/2012/03/document2008-12-18-131845-2.pdf.

352 "Bouquets & Brickbats," *Gambit*, December 22, 2008.

353 Letter from John R. Crane to Mary L. Landrieu, September 17, 2009. http://levees.org/wp-content/uploads/2009/09/2009_us_IG-Response-to-Landrieu2.pdf.

Chapter 8: Facing Off with the Army Corps

354 Email from the author to Jon Donley, February 10, 2009. http://levees.org/2/wp-content/uploads/2019/01/Email-from-author-to-Jon-Donley-2-10-2009.pdf.

355 Jed Horne, *Breach of Faith: Hurricane Katrina and the Near Death of a Great American City* (New York: Random House, 2006).

356 Email from Jed Horne to the author, February 28, 2009. http://levees.org/2/wp-content/uploads/2019/01/Email-Jed-Horne-to-author-2-28-2009.pdf.

357 Sandy Rosenthal, Levees.org Book Review: "Breach of Faith: Hurricane Katrina and the Near Death of a Great American City," *Market Wire*, January 12, 2010.

358 Email from Milena Merrill to the author, February 14, 2009. http://levees.org/2/wp-content/uploads/2019/01/Email-from-Milena-Merrill-to-author-2-14-2009.pdf.

359 Email from James O'Byrne to the author, April 15, 2009. http://levees.org/2/wp-content/uploads/2019/01/Email-James-OByrne-to-author-4-15-2009.pdf.

360 Screenshot: Outreach Process Partners webpage. http://levees.org/2/wp-content/uploads/2018/01/OPP-top-half-webpage.jpg.

361 Screenshot: Outreach Process Partners webpage. http://levees.org/2/wp-content/uploads/2018/01/OPP-bottom-half-webpage.jpg.

362 FedSpending.org, "Contracts Search Results (FY 2007)." https://www.fedspending.org/fpds/fpds.php?datype=T&database=fpds&fiscal_year=2007&maj_agency_cat=97&stateCode=LA&detail=-1&pop_cd=LA02&sum_expand=CA.

363 FedSpending.org, "Contracts Search Results (FY 2008)." https://www.fedspending.org/fpds/fpds.php?stateCode=LA&pop_cd=LA02&sortp=r&maj_agency_cat=97&sum_expand=CA&detail=-1&datype=T&reptype=r&database=fpds&fiscal_year=2008&submit=GO.

364 Email from the author to Levees.org supporters, May 5, 2009. http://levees.org/2/wp-content/uploads/2019/01/Email-author-to-Levees.org-supporters-5-5-2009.pdf.

365 Email from Janice Roper-Graham to Stacy Mendoza, May 5, 2009. http://levees.org/2/wp-content/uploads/2017/12/LeveesOrg-exposes-OPP.pdf.

366 Pia Malbran, "Outrage of Army Corps' $4.7M Contract," CBS News, May 8, 2009.

367 Georgianne Nienaber, "Army Corps of Engineers in New Orleans: Buying Advice or Spin?" *Huffington Post*, May 7, 2009.

368 Jonathan Carey Donley, "Affidavit of Personal Witness," State of Texas, County of Wilson County, June 9, 2009. http://levees.org/2/wp-content/uploads/2016/11/affidavit-Jon-Donley.pdf.

369 Email from the author to Levees.org supporters, June 18, 2009. http://levees.org/2/wp-content/uploads/2019/01/Email-author-to-Levees.org-supporters-6-18-2009.pdf.

370 Sandy Rosenthal, "Editor of Times Picayune replies to readers on Corps Internet Scandal," *Times-Picayune*, July 29, 2009.

371 Email from Jim Amoss to Levees.org supporters, July 28, 2009. http://levees.org/2/wp-content/uploads/2018/12/Email-Jim-Amoss-to-writers.pdf.

372 Letter from Levees.org to US Senator Mary Landrieu, June 22, 2009. http://levees.org/2/wp-content/uploads/2018/12/Letter-to-Senator-Landrieu-6-22-2009.pdf.

373 Blog post: Sandy Rosenthal, "Call to Senator Landrieu to request investigation of alleged Corps of Engineers Internet Scandal," June 23, 2009. http://levees.org/2009/06/23/call-to-senator-landrieu-to-request-investigation-of-alleged-corps-of-engineers-internet-scandal/.

374 Email from Aaron Saunders to Dominic Massa, June 23, 2009. http://levees.org/2/wp-content/uploads/2019/01/Email-Aaron-Saunders-to-Dominic-Mass-6-23-2009.pdf.

375 Eventually the *Times-Picayune* did cancel the *Levees News and Views* blog in September of 2018 citing changes to the website's infrastructure.

376 Mark Schleifstein, "Complaints about Corps of Engineers deserve investigation, Sen. Mary Landrieu says," *Times-Picayune*, August 6, 2009.

377 Letter from Mary L. Landrieu to Gordon S. Heddell, August 4, 2009. http://levees.org/wp-content/uploads/2009/09/Landrieu-Letter-to-DOD_IG-August-4.pdf.

378 Letter from John R. Crane to Mary L. Landrieu, September 17, 2009. http://levees.org/wp-content/uploads/2009/09/2009_us_IG-Response-to-Landrieu2.pdf.

379 "Jon Donley: The Corps is one of most powerful machines in the U.S.," December 15, 2009. http://levees.org/2009/12/15/jon-donley-the-corps-is-one-of-most-powerful-machines-in-the-u-s/.

380 Mark Schleifstein, "Department of Defense Inspector General closes investigation into allegations of derogatory postings to NOLA.com by Army Corps of Engineers employees," *Times-Picayune*, September 29, 2009.

381 Letter to the editor, "Levees.org responds to UNO engineering professor's Christmas Day letter," NOLA.com, December 29, 2008.

382 Letter to the editor, "Let's be kinder to the corps, for safety's sake," *Times-Picayune*, December 25, 2008.

383 William R. Freudenburg, et al., *Catastrophe in the Making: The Engineering of Katrina and the Disasters of Tomorrow* (Washington, DC: Island Press, 2009), 73.

384 Mark Schleifstein, "Hurricane Katrina flood ruling upheld by federal appeals court," *Times-Picayune*, March 3, 2012.

385 Ibid.

386 John Schwartz, "New Ruling on Katrina Favors Corps of Engineers," *New York Times*, September 24, 2012.

387 Letter from Alisa T. Henderson to the author, September 18, 2009. http://levees.org/wp-content/uploads/2011/03/FEMA-Letter-9-18-2009-1.pdf.

388 Boyd received his PhD in 2011.

389 Ezra Boyd, "Assessing the Benefits of Levees: An Economic Assessment of U.S. Counties with Levees," commissioned by Levees.org, December 23, 2009. http://levees.org/wp-content/uploads/2010/01/UsCountiesWithLeveesPaper_Boyd2.pdf.

390 Mark Schleifstein, "Levee statistics point up to their importance to the nation's economy," *Times-Picayune*, January 2, 2010.

391 "Fighting to fix flood insurance." http://levees.org/2/wp-content/uploads/2019/01/Senator-Mary-Landrieu-uses-Levees.orgs-graphic.png.

392 Email from Marc Levitan to the author, June 17, 2008. http://levees.org/2/wp-content/uploads/2019/01/Email-Mark-Levitan-to-author-6-17-2008.pdf.

393 Letter from Nicole Hobson-Morris to the author, July 29, 2010. http://levees.org/wp-content/uploads/2010/08/NRHP-Notice-of-Receipt-of-Nomination-First-Draft.pdf.

394 Email from the author to Kevin McGill, February 2, 2011. http://levees.org/2/wp-content/uploads/2019/01/Email-author-to-Kevin-McGill-2-2-2011.pdf.

395 Kevin McGill, "Hurricane Katrina levees: Group wants places where levees failed added to national register," Associated Press, February 7, 2011.

396 Tammy Conforti, "USACE National Levee Safety Program," PowerPoint presented at Association of Flood Plain Managers annual conference in Reno-Sparks, Nevada, May 2008.

397 Associated Press, "122 levees across the nation at risk of failing," February 2, 2007.

398 Congressional Research Service Report for Congress, Order Code RL33298, "Supplemental Appropriations: Iraq and Other International Activities; Additional Hurricane Katrina Relief," June 9, 2006.

399 United States Army Corps of Engineers, Engineering Technical Letter, ETL 1110-2-575, "Engineering and Design Evaluation of I-Walls," September 1, 2011.

400 Ibid.

401 News release: United States Army Corps of Engineers, Public Affairs Office, "Miss., Atchafalaya inspections begin Friday," October 8, 2006.

402 Conforti, "USACE National Levee Safety Program."

403 US Army Corps of Engineers, "Periodic Inspection Report 9 Update," presented to Dallas City Council, Dallas, Texas, June 3, 2009.

404 Public Law 110-114, title IX, §9004, November 8, 2007, 121 Stat. 1288.

405 Public Law 109-308, 120 STAT. 1725, October 6, 2006. http://levees.org/2/wp-content/uploads/2017/12/'Pets-Evacuation-and-Transportation-Standards-Act-of-2006.pdf.

406 Nomination to NRHP of March 25, 1911, Factory Triangle Shirtwaist Factory fire in New York City. https://npgallery.nps.gov/NRHP/GetAsset/05958ef0-fc22-429d-b350-b1372a70ddd7?branding=NRHP.

407 Sandy Rosenthal, "Letters in Support of Levees.org's Nomination of Breach Sites to National Register of Historic Places." https://www.scribd.com/lists/3190704/Letters-in-Support-of-Levees-org-s-Nomination-of-Breach-Sites-to-National-Register-of-Historic-Places.

408 Email from the author to Levees.org supporters, August 16, 2011. http://levees.org/2/wp-content/uploads/2019/01/Email-author-to-Levees.org-supporters-8-16-2011.pdf.

409 Email from Pat Duncan to the author, August 17, 2011. http://levees.org/2/wp-content/uploads/2019/01/Email-Pat-Duncan-to-author-8-17-2011.pdf.

410 Letter from Michael Fantaci of LeBlanc Butler LLC Attorneys at Law, October 20, 2010. https://www.scribd.com/document/62377071/LeBlanc-Butler-Law-Offices-Legal-Opinion.

411 Letter from Bradley Vogel to Nicole Hobson-Morris, August 11, 2011. https://www.scribd.com/document/62544613/National-Trust-for-Historic-Preservation.

412 John Sellers, "AP Releases Style Guide for 9/11," *Reuters*, August 10, 2011.

413 "Accused 9/11 plotter Khalid Sheikh Mohammed faces New York trial," CNN, November 13, 2009. http://edition.cnn.com/2009/CRIME/11/13/khalid.sheikh.mohammed/index.html.

414 "Tropical Cyclone Report, Hurricane Katrina," National Hurricane Center, December 20, 2005, 11. http://levees.org/2/wp-content/uploads/2018/08/Tropical-Cycline-Report-Hurricane-Katrina-UPDATED-14-Sept-2011.pdf.

415 Transcript: "Louisiana National Register Review Committee Meeting, Capitol Park Welcome Center, Baton Rouge, Louisiana," November 17, 2011. https://www.scribd.com/document/76963272/Transcript-of-November-17-2011-Meeting-Minutes.

416 Letter from Pam Breaux to Terrence Salt, December 27, 2011. https://www.scribd.com/document/76962530/Cover-letter-Pam-Breaux-to-Terrence-Salt-Dec-27-2011.

417 Kevin McGill, "Katrina levee breach won't get historic listing," Associated Press, June 15, 2012.

418 Letter from Carol D. Shull to the author, June 14, 2012. http://levees.org/2/wp-content/uploads/2017/12/Carol-Shull-Interim-Keeper-NRHP.pdf.

419 Alexandra Lord, "Writing national significance for National Historic Landmarks" (webinar), August 21, 2013.

420 Letter from Carol D. Shull to the author, June 14, 2012. http://levees.org/2/wp-content/uploads/2017/12/Carol-Shull-Interim-Keeper-NRHP.pdf.

421 Ibid.

422 Bob Marshall, "N.O. levee inspections fell short of federal mandate," *Times-Picayune*, December 5, 2005.

423 Email from Robert G. Bea to the author, March 6, 2012. http://levees.org/2/wp-content/uploads/2019/01/Email-Robert-Beat-to-author-3-6-2012.pdf.

424 Email from Oliver Houck to the author, March 8, 2012. http://levees.org/2/wp-content/uploads/2019/01/Email-Oliver-Houck-to-author-3-8-2012.pdf.

425 Photograph: http://levees.org/2/wp-content/uploads/2015/05/9862266683_8b36f20ab1_k.jpg.

426 Mark Schleifstein, "Study: Corps decisions, not Orleans Levee Board, doomed canal walls in Katrina," *Times-Picayune*, August 7, 2015.

427 Emails from Tim Doody to the author, September 14, 2015 to December 14, 2015. http://levees.org/2/wp-content/uploads/2019/01/Email-Tim-Doody-to-author-9-14-15-to-12-14-15.pdf.

428 Engineering Regulation No. 1130-2-530, "Flood Control Operations and Maintenance Policies," issued October 30, 1996. http://www.publications.usace.army.mil/Portals/76/Publications/EngineerRegulations/ER_1130-2-530.pdf.

429 Sandy Rosenthal and H.J. Bosworth Jr., "Closer eye on levees after Katrina," *Times-Picayune*, May 22, 2012.

430 Email from the author to Tim Doody, September 27, 2011. http://levees.org/2/wp-content/uploads/2019/01/Email-exchange-Tim-Doody-and-author-9-27-2011.pdf.

431 Email from Tim Doody to the author, September 27, 2011. http://levees.org/2/wp-content/uploads/2019/01/Email-exchange-Tim-Doody-and-author-9-27-2011.pdf.

432 Email from John Barry to the author, December 29, 2011. http://levees.org/2/wp-content/uploads/2019/01/Email-John-Barry-to-author-12-29-2011.pdf.

433

Ever," *New York Times Magazine*, October 2, 2014.

434 Ibid.

435 Douglas Woolley and Leonard Shabman, "Decision-Making Chronology for the Lake Pontchartrain & Vicinity Hurricane Protection Project, Submitted to the Institute for Water Resources of the U.S. Army Corps of Engineers," March 2008, 4–19. http://levees.org/wp-content/uploads/2010/07/Woolley-Shabman-Study.pdf.

436 Ibid., 4-2.

437 "Interagency Performance Evaluation Taskforce," Volume 3, US Army Corps of Engineers, 28.

438 Woolley and Shabman, "Decision-Making Chronology for the Lake Pontchartrain & Vicinity Hurricane Protection Project."

439 Ibid., 4–27.

440 "Interagency Performance Evaluation Taskforce," Volume 3, US Army Corps of Engineers, 28.

441 C. Andersen, J. Battjes, D. Daniel, and B. Edge, "The New Orleans Hurricane Protection System: What Went Wrong and Why: A Report by the American Society of Civil Engineers External Review Panel," American Society of Civil Engineers, Reston, Virginia, 2007, ES-5. http://levees.org/wp-content/uploads/2010/06/American-Society-of-Civil-Engineers-What-Went-Wrong-ERPreport-1.pdf.

442 The final version of the Decision-Making Chronology was published in March 2008.

Chapter 9: Bayoneting the Wounded

443 Comment by Moderatel on article by Bruce Alpert, "Jindal aide: Army Corps of Engineers is a 'complete disaster,'" *Times-Picayune*, February 7, 2013. http://levees.org/2/wp-content/uploads/2017/12/Corps-comments-part-3.pdf.

444 Email from Garret Graves to the author, April 17, 2006. http://levees.org/2/wp-content/uploads/2019/01/Email-Garret-Graves-to-author-4-17-2006.pdf.

445 Ibid.

446 Independent Levee Investigation Team, University of California, Berkeley, July 31, 2006, 15-4. http://projects.ce.berkeley.edu/neworleans/report/CH_15.pdf.

Chapter 10: A Major Coup

447 J. David Rogers, G. Paul Kemp, et al., "Interaction between the U.S. Army Corps of Engineers and the Orleans Levee Board preceding the drainage canal wall failures and catastrophic flooding of New Orleans in 2005," *Water Policy*, Volume 17, Issue 4, August 2015. http://wp.iwaponline.com/content/17/4/707#sec-3.

448 Bruce Alpert, "Jindal aide, Army Corps of Engineers is a 'complete disaster,'" *Times-Picayune*, February 7, 2013.

449 Senate Environmental and Public Works Committee hearing, February 7, 2013, 6–7. https://www.epw.senate.gov/public/index.cfm/hearings?ID=92B2B96F-0845-AF55-539C-B9590BD2E142.

450 Comment by Sandy Rosenthal to article by Bruce Alpert, "Jindal aide, Army Corps of Engineers is 'a complete disaster,'" *Times-Picayune*, February 7, 2013. http://levees.org/2/wp-content/uploads/2017/12/Corps-comments-part-3.pdf.

451 Comment by Moderatel to Sandy Rosenthal on article by Bruce Alpert, "Jindal aide, Army Corps of Engineers is 'a complete disaster,'" *Times-Picayune*, February 7, 2013. http://levees.org/2/wp-content/uploads/2017/12/Corps-comments-part-3.pdf.

452 Business Council of New Orleans and the River Region, "It's Time for Experts to Manage Levee Safety." http://levees.org/2/wp-content/uploads/2017/04/Ad-in-Times-Picayune-11-16-05.pdf.

453 Jim Tucker, "Press Release: Louisiana Republican Legislative Delegation," November 22, 2005. http://levees.org/2/wp-content/uploads/2017/04/LA-Republican-Legislative-Delegation.pdf.

454 Jed Horne, " 'Help us please,' " *Times-Picayune*, September 2, 2005. Part of the Hurricane Katrina coverage awarded the 2006 Pulitzer Prize for Public Service.

455 Bob Marshall, "Corps never pursued design doubts," *Times-Picayune*, December 30, 2005. Part of the Hurricane Katrina coverage awarded the 2006 Pulitzer Prize for Public Service.

456 Email from Robert Bea to H. J. Bosworth, Jr., March 13, 2013. http://levees.org/2/wp-content/uploads/2018/01/Email-chain-Dr-Bob-Bea-HJ-and-Sandy.pdf.

457 Mark Schleifstein, "U.S. Supreme Court lets stand ruling tossing Hurricane Katrina judgment against Army Corps of Engineers," *Times-Picayune*, June 24, 2013.

458 Email from Paul Kemp to the author, April 8, 2013. http://levees.org/2/wp-content/uploads/2019/01/Email-Paul-Kemp-to-author-4-8-2013.pdf.

459 Email from J. David Rogers to the author, April 10, 2013. http://levees.org/2/wp-content/uploads/2019/01/Email-David-Rogers-to-author-4-10-2013.pdf.

460 J. D. Rogers, "Development of the New Orleans Flood Protection System prior to Hurricane Katrina," *Journal of Geotechnical and Geoenvironmental Engineering*, ASCE, May 2008.

461 Email from J. David Rogers to the author, April 16, 2013. http://levees.org/2/wp-content/uploads/2019/01/Email-J.-David-Rogers-to-author-4-16-2013.pdf.

462 Staff report, "Limits to Development: When Should You Avoid Building at All Costs?" *NBC News*, March 29, 2014.

463 Ryan Sabalow and Dale Kasler, "Oroville Dam repairs aren't enough, feds warn. Should state be forced to plan for a mega-flood?" *Sacramento Bee*, November 1, 2018.

464 Mottomo Sushi. https://www.zmenu.com/mottomo-sushi-rolla-online-menu/.

465 Volume 1, *Interagency Performance Evaluation Taskforce*, US Army Corps of Engineers, 44.

466 Ibid., 28.

467 Ivor van Heerden, G. Paul Kemp et al, "Team Louisiana," Louisiana Department of Transportation and Development, December 18, 2006, 141.

468 Email from J. David Rogers to the author, May 10, 2013. http://levees.org/2/wp-content/uploads/2019/01/Email-J.-David-Rogers-to-author-5-10-2013.pdf.

469 Stephen Braun and Ralph Vartabedian, "Levees Weakened as New Orleans Board, Federal Engineers Feuded," *Los Angeles Times*, December 25, 2005.

470 Email from Bruce Feingerts to the author, April 26, 2013. http://levees.org/2/wp-content/uploads/2019/01/Email-Bruce-Feingerts-to-author-3-26-2013.pdf.

471 Douglas Woolley and Leonard Shabman, "Decision-Making Chronology for the Lake Pontchartrain & Vicinity Hurricane Protection Project, Submitted to the Institute for Water Resources of the U.S. Army Corps of Engineers," March 2008, 2–48. http://levees.org/wp-content/uploads/2010/07/Woolley-Shabman-Study.pdf.

472 Rogers, Kemp, et al., "Interaction between the US Army Corps of Engineers and the Orleans Levee Board."

473 Ibid.

474 "Minutes of the Special Board Meeting of the Board of Commissioners of the Orleans Levee District," May 30, 1991, 8. http://levees.org/2/wp-content/uploads/2018/05/May-30-1991-Special-OLB-Meeting-Minutes.pdf.

475 "Minutes of Airport Committee Meeting held on November 7, 1990," US Army Corps of Engineers Presentation on Flood Protection for London Avenue Canal, 3. http://levees.org/2/wp-content/uploads/2018/12/1990-11-07-Committee-Minutes.pdf.

476 Email from J. David Rogers to the author, September 5, 2013. http://levees.org/2/wp-content/uploads/2019/01/Email-J.-David-Rogert-to-author-9-5-2013.pdf.

477 Email from Jerome Delli Priscoli to H. J. Bosworth, Jr., January 4, 2014. http://levees.org/2/wp-content/uploads/2019/01/Email-Jerome-Delli-Priscoli-to-HJ-Bosworth-Jr-1-4-2014.pdf.

478 Ibid.

479 Ibid.

480 Ibid.

481 J. David Rogers, G. Paul Kemp, et al., "Did the Local New Orleans Levee Board Drown its City?" manuscript no. WPOL-D-13-00149, submitted to *Water Policy*, September 20, 2013.

482 Email from the author to J. David Rogers, February 28, 2014. http://levees.org/2/wp-content/uploads/2019/01/Email-author-to-J.-David-Rogers-and-Raymond-Seed-2-28-2014.pdf.

483 Email from the author to J. David Rogers and Raymond Seed, April 5, 2014. http://levees.org/2/wp-content/uploads/2019/01/Email-author-to-J.-David-Rogers-and-Raymond-Seed-4-5-2014.pdf.

484 Email from Raymond Seed to the author, April 6, 2014. http://levees.org/2/wp-content/uploads/2019/01/Email-Raymond-Seed-to-author-4-6-2014.pdf.

485 Email from Raymond Seed to the author, April 11, 2014. http://levees.org/2/wp-content/uploads/2019/01/Email-Raymond-Seed-to-author-4-11-2014.pdf.

486 Draft: "Interaction between the U.S. Army Corps of Engineers and the Orleans Levee Board Preceding the Drainage Canal Wall Failures and Catastrophic Flooding of New Orleans in 2005," *Water Policy*. http://levees.org/2/wp-content/uploads/2018/02/Paper-submittd-to-WPOL-April-2014.pdf.

487 Email from Jerome Delli Priscoli to H. J. Bosworth, Jr., September 21, 2014. http://levees.org/2/wp-content/uploads/2019/01/Email-Jerome-Delli-Priscoli-to-HJ-Bosworth-Jr.-9-21-2014.pdf.

488 Email from the author to J. David Rogers, September 22, 2014. http://levees.org/2/wp-content/uploads/2019/01/Email-author-to-J.-David-Rogers-9-22-2014-1.pdf.

489 Email from H. J. Bosworth, Jr., to the author, September 23, 2014. http://levees.org/2/wp-content/uploads/2019/01/Email-H.J.-Bosworth-to-author-9-23-2014.pdf.

490 Ibid.

491 Email from Jerome Delli Priscoli to H. J. Bosworth, Jr., November 4, 2014. http://levees.org/2/wp-content/uploads/2019/01/Email-from-Jerome-Delli-Priscoli-to-HJ-Bosworth-Jr-11-4-2014.pdf.

492 Email from Cheron Brylski to the author, November 14, 2014. http://levees.org/2/wp-content/uploads/2019/01/Email-Cheron-Brylski-to-author-11-14-2014.pdf.

493 Rogers, Kemp, et al., "Interaction between the US Army Corps of Engineers and the Orleans Levee Board."

494 Campbell Robertson and John Schwartz, "Decade After Katrina, Pointing Finger More Firmly at Army Corps," *New York Times*, May 24, 2015.

495 Mark Schleifstein, "Study: Corps decisions, not Orleans Levee Board, doomed canal walls in Katrina," *Times-Picayune*, August 7, 2015.

496 Ibid.

497 Ibid.

498 Peter Ehrhard, "New Orleans' levee system failure after Katrina has mistaken culprit," Phys.org. https://phys.org/news/2015-05-orleans-levee-failure-katrina-mistaken.html.

499 "Flood damage after Hurricane Katrina could have been prevented, experts say," *ScienceDaily*, August 24, 2015.

500 John Burnett, "Billions Spent On Flood Barriers, But New Orleans Still A 'Fishbowl,'" NPR, August 28, 2015.

501 Cain Burdeau, "Katrina marker commemorates flooding of Lower 9th Ward," Associated Press, March 15, 2016.

502 Email from Om Dixit to H. J. Bosworth, Jr., July 6, 2016. http://levees.org/2/wp-content/uploads/2019/01/Email-Om-Dixit-to-HJ-Bosworth-Jr-7-6-2016.pdf.

503 Blog post: Len Bahr, "Former Orleans Levee Board exonerated of culpability for Katrina flooding," July 10, 2015. http://lacoastpost.com/blog/?p=49255.

mango
PUBLISHING GROUP

Printed in the USA
CPSIA information can be obtained
at www.ICGtesting.com
JSHW030217290524
63780JS00009B/12

9 781642 503272